She headed for the Commander's office, wondering what that was so important, and knocked on the door.

"Enter," the usual response.

"You wanted to see me, sir?"

"Have a seat, Cabrillo. Close the door. I have some news for you regarding a new assignment."

"What's that, Sir?"

"You've received new PCS orders."

"What? But Commander, I've only been assigned here for a few months. What's up with that, sir?"

"I don't know where they're coming from or who ordered it, Ali. The orders arrived earlier today. All I can tell you is I got a call from AirPac this morning confirming your transfer. You are to proceed immediately to your new squadron, currently deployed on USS Guantanamo Bay, deployed in the Med."

"What the hell! How's that even possible, sir?"

"Just proceed as soon as possible. I hate to lose you, though. You're the best instructor we have. I'm not even sure I'll get a replacement for you."

"Thanks for that, Commander. I'll miss everyone here, too."

"Okay, you have your orders. Check out with admin; They have your orders and records ready. Shove off, and good luck."

"Yes sir, good luck to you too, Commander."

Walking out, collecting her gear and orders, she wondered what was going on. "What's this about? It's good I don't have an apartment with a lease!" She made the rounds, saying her goodbyes to everyone, and that was it. She detached the command and was off.

In Transit

The contract airliner carrying Lieutenant-Commander Althea 'Ali' Cabrillo landed at picturesque NAS Sigonella, Sicily. She'd been traveling on PCS orders on the flight—one of many flying between Navy and Air Force bases worldwide, in transit to her new command.

It glided in on the final approach to NAS Sigonella with a smooth flare, lightly touched down, and rolled to a stop in front of the terminal. Air stairs rolled up to it, the doors swung open, and passengers started deplaning.

Ali Cabrillo strode down the steps, noticing Mount Aetna looming large in the distance. She stopped and did a wide-eyed, open-mouthed double-take at the striking sight.

After a short, incredulous moment marveling at it, Ali slow-walked it towards the terminal building, still gawking at the snow-capped smoking volcano off in the distance. Quite a sight for a Texas girl! She'd never seen a volcano, let alone an active one.

Her new assignment was as executive officer of an MH60-R Seahawk helicopter squadron detachment, deployed on the USS Guantanamo Bay, an LHA-type amphibious assault ship. The ship was at sea, so Ali and the few others heading for it had to fly into the air station at Sigonella and wait for a COD flight to the ship for a few hours.

After the long flight from Oceana, she planned on taking a break and resting at the air terminal cafeteria, waiting for the COD to the ship. Dressed in her best Dress Blues, polished brass, a four-row ribbon rack, and black patent leather shoes, Ali headed for the baggage counter to collect her gear. A familiar voice called out from behind, "Ali? Ali Cabrillo from Pensacola?"

She turned to see if it was him. Yes, it was Sam, one of her flight school classmates with whom she'd had a brief amorous fling, Major Samuel 'Sammy' Nash, a Marine Aviator himself. She grins a little and thinks to herself, "Oh shit, what's he doing here?"

"Hey, Sam, what are you doing here? It's been a long time, dude! When did you get here? Have you been waiting very long?"

"Great to see you again, too, Ali! I flew in yesterday to pick up my new squadron."

"What a coincidence—I'm doing the same thing! They've assigned me as the new XO to a Seahawk Det. Yeah, it's been a while, for sure!"

Sam sighed wistfully and blurted out, "Yes, it has! Twelve years, to be exact. Too long, Ali! I've missed you a lot!"

Sam always had a thing for Ali. Who wouldn't? Ali was a striking woman with dark flowing hair, a light olive Italian or Spanish-looking complexion, and big, beautiful dark eyes.

Ali wasn't a tall woman, though; most Aviators aren't—easy blackouts during high G maneuvers for a tall pilot—something about the blood flow distance from the heart to the brain and that it takes longer, causing the blackouts.

Flattered but not wanting to show it, she tried to keep the conversation from getting serious. "Sam, you're such a schmooze! You knew that flight school thing wasn't going anywhere. We didn't have time for it then or now, doing what we do for a living."

"Yeah, I know; I still think about you a lot, though."

She smiled, then quickly changed the subject. "Well, that's sweet of you anyway, Sam, but stow it. Not happening! What have you been doing since flight school? What are you flying now?"

Sam threw her a smirking grin, screwing with her a little, bragging it up as Marines like to do. He knew she wanted to fly Super Hornets and would be jealous as a cat to find out what he was doing.

"F-35s! After flight school, I got assigned to a harrier unit. Later, they started transitioning to F-35s. From there, I got assigned to an RAG outfit for training on the B model. So now, here I am, flying them!"

His bragging pissed her off a little. She tilted her head, pursed her lips together in a jealous frown, flashed him the stink-eye, and sarcastically quipped,

"Well, shit, Sam! You always were a lucky dude! I didn't get to fly Super Hornets. I'm flying Seahawks now. It's been helicopters for me since flight school. Congrats on your promotions, too! Last time I saw you, you were still a second Lewy."

Sam beamed and congratulated himself for the jab and her reaction, then boasted a little more.

"Yeah, the Corps has been good for me. How about you?"

"Oh, it's been okay. I love flying. That makes it worthwhile, even if I didn't get fighters. Maybe I can finagle that in the future."

Ali changed the subject to escape the everlasting bragging. "What time is the COD flight to the ship due?"

"I'm not sure. The PAX counter told me it was supposed to be here in a couple of hours, but that might change."

"When did you say you got here, Sammy?"

"Yesterday. We were late getting in, so I missed the COD flight to the ship, or it never made it here, or whatever, I'm not sure."

"Where are you coming from?"

"I left Cherry Point the day before yesterday. I've been here since then, waiting for a ride to the ship, enjoying the view and the nice weather."

"Too bad we can't get off base, get dinner somewhere, and catch up properly. I hear they have some pretty good Italian food here." They both laugh a little at her joke.

"I'm sure they do, Ali; it's Sicily, but you know how these flight schedules go. You can never be sure when they'll show up. One time, I caught a contract flight out of Travis, heading to Futenma in Okinawa, and had to wait overnight in the terminal at Travis for it. That flight was so loaded with dependents and kids that I would never have gotten a seat without orders!"

"Yeah, hurry up and wait. I know the drill!" They sat together in the air terminal's lounge, drinking coffee, chatting, and catching up, waiting for their flight. When it finally showed up, the PAX counter announced, "All passengers for USS Gitmo, COD flight will board in 15 minutes. Please proceed to the check-in counter for boarding." They loaded into the Osprey, took off, and headed for the ship.

Ali was a little uneasy about their ride. "One of those Osprey tilt-rotor accidents waiting to happen!" She wasn't fond of them. "Their safety record isn't the best, but hey, it's a ride, I guess."

Flights on Osprey's differ from the old C-2 greyhound COD aircraft. They're VSTOL, 'Vertical or Short Take-Off and Landing' that don't need any runway distance to get off the ground. Since this flight was from a regular airfield, it still had to use the standard traffic pattern and tower procedures. Otherwise, it would have been able to take off straight from the Air Terminal ramp.

After a couple of hours of flying time to the ship, the Osprey made its approach. All aircraft on these kinds of ships don't land like they do on a big deck carrier. There's no arresting cable to grab and stop fast-moving aircraft careening down the deck.

They approached the ship from the port (left) side and landed on designated spots, side slipping to starboard (right). The LSE Petty officer then waved them onto the deck with hand signals.

The Osprey descended slowly and touched down with a light thud on the deck. After it shut down, Cabrillo, Nash, and the others unstrapped their harnesses, stood up from the uncomfortable webbed jump seats, climbed out on the deck, and headed for the catwalks to the access ladders to the next lower deck.

The Gitmo would be their home for the next few months—if they were lucky, that's all it would be. Overseas deployments like this often get extended, and it's home for even longer—sometimes much longer! Members of each command met the new arrivals, escorting them to their admin offices to check in. Ali was the exception; her new CO, Commander Scott Deering, met her.

Deering was a medium-height fellow with a stocky build and piercing blue eyes. At first, his intense gaze and stiff demeanor slightly intimidated Ali, but he was welcoming and pleasant to her.

"Welcome aboard, Ms. Cabrillo! Pleased to meet you finally. We've been waiting for you. How was your trip out from San Diego?"

"Good afternoon, sir. I'm pleased to meet you as well. It was a long trip but pleasant. We had a couple of layovers before we arrived at Sigonella, and then we had to wait there for the COD to the ship."

Ali was relieved that he was friendlier than he looked. "I ran into an old friend there. We spent the wait time catching up." She added.

"Ah yes, Major Nash. You two went to flight school together, didn't you?"

"Yes sir, we did, but how did you know?"

"Nash's new CO, Lieutenant Colonel Bradford, and I are buds. When I told him about you, he mentioned that Nash and you were classmates. He was one of your instructors, wasn't he?"

"Ah yes, I remember him well, sir." Ali remembered Bradford very well; "So, that prick Bradford's here too! Old Iron ass! Had to happen sometime, I guess—the Navy's a small world, after all."

Deering continued, "I don't know how he knew you both were getting assigned to squadrons on the same ship simultaneously, but he did. I never asked him that."

"Hmm, that's interesting. We had a good time catching up and waiting for the COD. I have to admit, Mount Aetna impressed me a lot. I'd never seen an actual volcano up close!"

Trying to make small talk, Deering launched into a lengthy diatribe on his experiences with volcanoes. "Yeah, it's pretty impressive and also disturbing! I remember Mount Pinatubo, in the Philippines. My Dad was stationed at Cubi Point when it erupted. I have a healthy respect for volcanos!" He went on and on and told her about his experiences there as a boy living in Binictican's dependent housing.

After he finished rambling about it, he let her get on with her check-in. "Well, let's get you settled in. Yeoman Perez will show you around and set you up with your stateroom." Perez responded, "Yes, sir, I'll take care of it."

"We'll get you started on pass-downs and review everything with Fox at squadron muster tomorrow morning," Deering added as Ali and Perez walked away.

"Thank you, sir. Pleasure meeting you."

"I'll take care of your paperwork, Ma'am. You can leave everything with me. Follow me, and I'll show you to your stateroom," Parez said as instructed.

"Thank you, Petty Officer Perez. Lead the way." Perez showed her to her new stateroom and got her settled in. It'd been a long, tiresome trip, so Ali crashed and tried to unwind. Duty would call soon enough—time now for some rack time.

Muster on Station

0700 the following day, squadron staff assembled on the hangar deck for muster (a roll call and headcount) to ensure everyone was present or accounted for, personnel inspections, and any administrative matters or announcements. That day was for Ali Cabrillo's introduction as the new Executive Officer.

Master Chief Banks bellowed, "Squadron attention on Deck!" as Deering, Cabrillo, and Fox approached. Banks turned to face Deering and delivered his muster report; "Squadron all present or accounted for, sir!"

"Very well, Master Chief," Deering responded. Typically, he wouldn't have been present for this. Normally, the squadron's Chief Petty Officer or a Senior First-Class Petty Officer handled musters themselves. Today, Deering was there to introduce Ali officially to everyone.

"Listen up, people. I want to introduce Lieutenant-Commander Cabrillo today. She's taking over as the executive officer from Lieutenant-Commander Fox and as the senior Aviator after me. Show her appropriate courtesies and respect. Mr. Fox and Ms. Cabrillo, please follow me to the wardroom. Master Chief, take charge."

"Yes, sir! Company dismissed. Turn to!" At that, everyone broke formation and headed off to their work areas. Banks overheard one of them spouting off about Ali already: "Wow, she's one hot tamale."

Banks scowled and barked at him, "Shit-can that kind of talk, sailor! She's your XO. Show her the proper respect! I don't want to hear that kind of scuttlebutt again - understand!"

"Okay, Master Chief, sorry. But you have to admit, she's pretty hot!"

"Your ass will be pretty hot, and you'll be even sorrier if she hears about it–now get your ass in gear and turn to!"

Banks had to admit she was a striking woman, but he didn't want his crew talking smack about her.

The Wardroom

The officers' wardroom on the GITMO resembled the dining room of a well-appointed hotel. It featured buffet-style service with dinnerware embossed with the ship's name, good-quality silver and crystal glassware, and enlisted personnel serving the officers like waiters on mess duty.

Breakfast in the wardroom always included pastries, made-to-order eggs, a buffet-style hot breakfast, and fresh coffee all the time. Rank has its privileges. Cabrillo, Fox, and Deering had breakfast and discussed squadron matters and the handover from Fox to Cabrillo.

"Well, Ms. Cabrillo, what do you think of the ship so far?"

"Very nice, sir, very well appointed." She couldn't say what she was thinking. "I'll bet the crew doesn't dine so well as this."

Ali was blue-collar that way and pro 'little guy' on the inside. She identified more with the enlisted people than the privileged officer class. She grew up in a family of modest means. To her, this seemed an unnecessary extravagance for military people.

Deering continued: "Mr. Fox plans to seek other opportunities outside the Navy."

Ali was curious and asked. "Have you got any concrete plans yet, Fox?"

"Well, I've had several outside offers, but I haven't decided yet. My circumstances have changed a lot, anyway. I feel like it's time for a change."

Fox wasn't about to mention that he'd received a couple of unfavorable Fitness Reports and had gotten into trouble with one of his female Aviators. Ali would learn about that issue soon. Deering shot a dirty look at Fox for bringing up his circumstances. Fox picked up the queue and didn't elaborate any further.

When an officer at any level, especially an Aviator of his or her rank and seniority, gets a bad fitness report, that's pretty much it. They're unlikely to be considered for any future promotions or choice assignments. So, he just decided, "What the hell, I'll just get a gig with one of the Navy contractors and put all this crap behind me!"

Fox continued, "Good pay, good hours, and home every night. Commercial test pilot, maybe." The chit-chat was mainly over, so Deering took his leave and headed for his office.

"Well, I'll leave you two to your pass-down duties. Ms. Cabrillo, again, welcome aboard. We're happy to have you with the squadron." "Yes, sir. Thank you, sir. I'm pleased to be here as well."

"Fox, start your checkout after you finish up with Ms. Cabrillo. You leave on the COD fight for Sigonella the day after tomorrow."

"Yes, sir, I'll get Lieutenant-Commander Cabrillo up to speed as quickly as possible, Sir."

"Very well then, carry on."

Ali and Fox both chimed in with, "Aye-aye, sir!" Deering was off. Fox's chilly relationship with Deering surprised Ali a bit. "What the hell was that all about?" She wondered, although her better judgment kept her from asking.

An Unexpected Coincidence

Cabrillo and Fox headed for the XO's office and spent most of that day reviewing her duties and receiving a thorough pass down from him. Ali felt he was rather cold and unfriendly. She realized he wanted to finish the handover quickly and leave the ship as soon as possible. Not knowing she would find the reason for this behavior soon.

After leaving the squadron XO's office, she headed to her stateroom to rest. She went down the passageway through Officer Country to her stateroom. As she arrived, Ali heard a giddy, excited little voice call from behind, "Ali? Is that you?"

Ali recognized the voice immediately; it was her old friend and partner in crime. She turned around to see Lieutenant Michelle Robbins, her bestie.

"Shelly! What are you doing here? OMG, what a surprise!"

"We're squadron mates! I heard you would be our new XO and wanted to say hi and welcome you. I've been with the squadron for a few months. Where are you coming from?"

"I was in a RAG outfit at North Island (replacement air group,) an advanced training squadron on a temporary assignment. I got PCS orders for my new unit, so now I'm here."

Curious, Ali probed Michelle, like an XO would, "How come you weren't at muster this morning, Lieutenant?"

"Oh, I was in medical; I wasn't feeling well. Ali, I'm so happy to see you again!"

"Yeah, it's great seeing you again, too, Shelly! I'm heading to my stateroom. Where's yours?"

"We're neighbors. I'm in the stateroom next to yours!"

"Cool! That's great, Shelly, but I'm XO now, your boss, so keep that in mind, okay?" Ali felt like she had to remind Michelle of that. She knew Michelle could get a bit too informal, a little too 'touchy-feely' for her own good, often at the wrong times.

Michelle took it in stride, though. With a cheery grin, she blurted out, "Yes, Ma'am!" and snapped a salute jokingly. After so long, Michelle was just overjoyed to see her, her closest and best friend ever, and her secret 'girl crush' again.

Ali replied: "Serious, Shelly, just keep it business around others, okay, especially Commander Deering. I don't know how to take him just yet."

"Yeah, I understand. He's pretty by the book. I got it. I'm cool with that; see you at Chow, maybe?"

"Sure, I'll be there. See you then, and don't call me by my first name where anyone else can hear you!"

"Okay, understood, XO."

A Personnel Issue

Ali was sitting in her office several days later, working on squadron duty assignments when the desk phone rang. "Strange. Who'd be calling me here, anyway?" she thought to herself.

It was Deering calling, "XO, please come to my office. I have an issue to discuss with you."

"Yes, sir, I'll be there right away."

She wondered what was up now. "Okay, wtf! What could that be—Shit, problems already? I just got here!" Aggravated, she might get chewed out for something she didn't know yet. "Well, I guess I'll find out shortly," Ali mutters to herself.

She was off to Deering's office, just a few steps from hers. Ali knocked on his door with the traditional three knocks. "Enter!" bellowed Deering. "Good afternoon, XO. Please, have a seat."

"You wanted to talk to me, Commander?"

"XO, I need to brief you regarding an issue involving Lieutenant Robbin's job performance and a recent disciplinary action I had to take with her."

Ali was stupefied but not surprised, considering it involved Michelle. "What's the issue, sir?"

"Well, as you know, you replaced Lieutenant-Commander Fox, but you're unaware of why. Long story short, Fox and Lieutenant Robbins were having an affair."

"She's also had performance issues before that. Then, I started receiving reports about their conduct from other squadron staff. What brought it to a head was when Captain Manning, the Ship's CO, saw them on liberty in Naples in what he described as a 'Public Display of affection.'"

"As you know, fraternization's frowned upon and strictly forbidden by Navy Regs, especially when they're part of the same unit."

"I see, sir. So, what happened then?"

"When I confronted Fox about it, he said she'd taken a shine to him, innocently, at first. Only later was it apparent that she'd hoped he would overlook some of her performance problems because of their relationship."

"How did you find out that was her motivation, sir?"

"I confronted her. She spilled her guts about it. I suppose she was hoping I'd be lenient. She told a different version of the story, though. She said Fox was pressuring her for sex because of her unsatisfactory fitness reports."

"Lieutenant Robbins claimed he told her he would overlook them if she agreed to an affair. I couldn't overlook that, especially since the captain saw them together and was aware of it, too."

"What did you do, Sir?"

"I was aware Fox was having problems at home. His marriage was falling apart, and his performance had also gone downhill. I assume, for the same reason. Even before this happened, I had to give him a marginal Fitness Report."

"When I confronted him with her accusations, He admitted they'd had an affair but claimed it was consensual between them. So, after that, I gave him a choice: resign his commission and deal with his problems without the pressure of duty and deployments or face an article 134 wrongful fraternization charge."

"What about Lieutenant Robbins? What did you do about her?"

"I put her and Fox both in hack for a month. Robbins is an excellent pilot, but she has some quirky personality and attitude problems. I gave her the benefit of the doubt, though, since Fox was her senior officer. I felt she needed someone to jack her up and motivate her to get back on track. In short, she needs someone like you to straighten her out."

"Yes, sir, I understand where you're coming from; I was one of her instructors at flight school. I know exactly what you mean."

"That's a big reason I asked for you. Since you were between permanent assignments, I requested you directly through ComNavAirLant."

"But how did you even know to request me in the first place, sir?"

"Well, as I mentioned when you checked in, Major Nash's new CO, Lieutenant Colonel Bradford, suggested it. He mentioned you had also been one of her instructors and worked with her through similar problems then."

"It was more like I had to put a boot in her ass, sir. She can be very inattentive and stubborn. We nearly washed her out because of it. However, the command decided we would work with her, and I was directed to work with her and straighten her out since she was an excellent pilot. So, I handled it. She responded, and it worked out."

"I got that from Bradford. Given what happened between her and Fox, I didn't think I could assign that task to another officer. That's why I requested you. That, and your outstanding record."

"Well, thanks for that, sir, for your confidence in me. I'll do what I can with her."

"That's all I can ask. If she doesn't improve, we'll have to take more drastic and punitive measures and ask her to resign!"

"I understand, Commander. I'll work with her. Would it be a problem to assign her as my co-pilot for now? I can give her more direct supervision that way."

"Do whatever you feel is required. It's entirely up to you. But keep track of what you do and document her outcomes."

"Understood, Commander. Was there anything else, sir?"

"That'll be all XO, dismissed."

"Aye-aye Sir."

As she left Deering's office, Ali thought, "Here we go again. What am I going to do with you, Michelle? Why are you such a twit?"

The situation made her reflect on what happened while she was Michelle's instructor at Jacksonville.

Back in the Day - Flight School

Ali and Michelle's backstory began with advanced flight training at the Norfolk training command. Ali was a Seahawk flight instructor, where she taught Michelle and other trainees the finer points of flying the machines. Despite their contrasting personalities, they worked well together.

Ali was a pragmatic, self-assured professional. Some might call her a know-it-all, since she always had opinions about everything and was right most of the time. On the other hand, Michelle was the easy-going type, opinionated, hypersensitive, and sometimes unsure of herself—undesirable traits for an Aviator in Ali's way of thinking.

Despite their personality differences, they became close friends. Ali often got frustrated with Michelle's flakey attitude. She thought Michelle was inattentive and made too many simple mistakes. Being the professional she was, Ali made extra efforts to correct what seemed like Michelle's dumb screw-ups over and over, which ended up with her giving "Attitude change" lectures.

Michelle never made mistakes flying, though. She was a top-notch Aviator in every way except for her quirky personality. Michelle often thought Ali was being a hard-ass and unfair to her, but she soon realized it was helping her a lot, or maybe she just changed for Ali.

After all the extra effort, time, and help Michelle got from Ali, she became very attached to her and an excellent pilot. Michelle was sure she wouldn't have made it through advanced training without Ali's direction and guidance. Ali liked Shelly but was uncomfortable with what she felt was too much personal attention Michelle heaped on her. Ali wasn't the best at picking up subtle queues!

"Damn it, Michelle, don't get so worked up; I'm trying to help you get this shit right! It's like sometimes you think I'm your mother or something. Straighten your ass out, or they'll wash you out!"

"I'm sorry! Please don't get pissed off at me. I appreciate all your help, that's all!"

"Just stick to business. Pay attention to what I tell you, and try harder, okay?"

"Okay, sorry, I mean yes, Ma'am."

After the conversation with Deering, Ali needed to have another serious chat with Lieutenant Robbins. "Lieutenant Robbins, report to the squadron XO's office." Michelle had to wonder, "WTF? Why so formal?" It never crossed her mind that it might be about her relationship with Fox and the Commander's solution. She knocked on the XO's office door, waiting for Ali to tell her to come in.

"Enter!" came the formal, business-like response from Lieutenant-Commander Cabrillo.

"Oh, shit! What did I do? You wanted to see me, Ali? I mean, Ma'am?"

"Yes, come in and sit down, please."

Now Michelle was nervous at Ali's tone.

"Is there anything wrong, Ma'am?"

Ali spoke as an XO would, "Not with me, but Commander Deering filled me in on your relationship with Lieutenant-Commander Fox and his actions against the both of you. He's ordered me to work with you and correct some of what he feels are your 'behavior and attitude' issues."

Michelle was upset and embarrassed that Ali had found out about her and Fox the way she had. "I'm sorry you had to hear that from him, Ali. I wanted to talk to you about it first, but I guess that opportunity has passed."

"Indeed, it has, Michelle! He told me he specifically requested that I take Fox's job because of it and our time during advanced training. He felt you would pay more attention if I worked with you. So, I'm here because of you, Michelle!"

"Damn! But how did he even know about that? I'm sorry if that's my fault, Ali. I never would have ever even known he would do that."

"He and Bradford are friends. Bradford filled him in on our history. He felt like I could straighten you out. He told me he wasn't comfortable bringing in another male officer to replace Fox because of you!"

"Why would he think that? Fox came on to me!"

"Well, I suppose he felt the same thing might happen again since Fox claimed you came on to him, and it was consensual after your bad Fitness Reports!" Ali spoke sarcastically to her now: "What else would you think his reasoning might have been, Michelle?"

"Oh, I see! That prick Fox blamed me for it, huh?" Michelle was livid that Fox would do that to her since she'd agreed to his advances. She paused momentarily, a little embarrassed, then said, "So, what now?"

"You will fly with me as my copilot. I'm to correct your performance deficiencies. So, start acting right, don't question orders, and above all, don't make smart-ass remarks to me or anyone else! That would be my first order and my friendly advice to you."

Ali tried to sound slightly more encouraged. "I know you have superior flying abilities, and so does Deering. That's the only reason he's giving you this shot. If you don't come around, he almost certainly would expect me to give you another adverse review."

"He would probably bring you to MAST again, maybe even a Court-Marshal, and try to wash you out with an administrative or BCD discharge! At the very least, it probably would also mean he'd put you in hack again for being a screw-up and not following orders!"

"Oh shit! Okay, I'll do my best, I promise! I hope you aren't angry with me, XO." Michelle made a point out of using her title.

"I'm not angry with you, Michelle. Maybe I'm frustrated that I'm in this position again, but that's how it goes. I'll be fair, but you'll have to straighten the fuck up, or we'll both be in trouble!"

"I'm also sure that you, I'm afraid, will be in much deeper shit than I will be, so quit with the screwing around! Okay, that's all, dismissed."

After their come-to-Jesus meeting, Michelle and Ali started flying together. Deering had paid close attention, noting Michelle's quick improvements, a testament to her efforts and sincerity in working with Lieutenant-Commander Cabrillo.

Days and weeks passed. Ali got into the swing of squadron life on deployment. The pace and workload increased dramatically as the conflict in Syria started getting hot rapidly. They flew as a team, Ali and Michelle as her copilot.

Many hours of operations, training flights, missions into Syria, and other 'un-named places' piled up. She and her crew flew several missions into the Syrian de-confliction zone almost daily, each a high-stakes operation moving marines, seals, and supplies to and from the op area. During these missions, they frequently took intense ground fire—a constant reminder of the dangers.

During this period in the Syrian civil war, the Navy and Marine Corps were increasingly involved in special operations missions against ISIS, primarily supporting the SDF and other coalition partners. Since they were flying the newest Seahawk, they flew more classified missions, ferrying the teams and other Special Operators into 'hot zones.'

A Faithful Mission

Off the coast of Syria, near Cypress, Cabrillo started early as she did most other mornings; up before reveille was piped through the ship, going over the squadron and the ship's Plan Of the Day, personnel issues, squadron operations schedules, aircraft maintenance issues, and other concerns an executive officer's usually responsible for.

This assignment as XO and senior pilot after the skipper surprised her. Ali had been an instructor pilot for so long that she thought it would be her entire career. She hadn't been enthusiastic about flying helicopters either. She was far more interested in flying Super Hornets. As many things fall out in the military, the Navy didn't see it that way.

When she completed advanced training, they needed Seahawk pilots, so that's where she ended up—more training to learn the ins and outs of flying them.

The ship and its complement of aircraft, helicopters, and combat marines had been supporting U.S. and coalition forces operating inside Syria against ISIS and Syrian government forces in the country's northeastern region. Today was a down day, a holiday routine day.

No flight ops were being flown, so she went up into the catwalks, off the port side forward, near the JP-5 refueling stations, to 'sight-see' out over the water.

Walking down the port-side passage, Nash noticed her heading out of the hatch to the catwalks and followed her to see what she was up to. "Hey Ali! I noticed you heading for the catwalks, so I thought I'd see what you were doing. Kind of out of your territory, isn't it?"

"Oh, hey, Sam. It is. When I get slack time, I come out here to watch the dolphins wake-surf in the ship's wake. It seems they think the ship's some big momma dolphin they can have fun with."

Nash laughed and said, trying to relate: "Ha, I never noticed that. Yeah, they seem to have fun enjoying themselves. Me, I'm looking forward to liberty in Haifa!"

Ali detected a wistful smile and timbre in his voice. She had to ask, "Why, Sam? You have a girlfriend there, do you?"

"Who me? Not really, no. I want to sight-see a little, kick back, and relax."

She laughed, knowing he was lying, and answered: "Don't bullshit me, Marine! I know you better than that." Sam looked down sheepishly and replied, "Well, I know several there." She laughed again: "I'll bet you do! Get out of here, you Jar-head skirt chaser!"

He decided he'd had enough abuse and said: "Okay, touché, you got me. I have to go. See you later." Amused at his reaction, she thought, "I'll never climb into the rack with him again! Probably got every disease known to man," giggling about it a little.

Very early the next day, 'zero dark thirty,' Ali was working squadron personnel training schedules. Michelle unlatched and cracked open the cabin door separating their adjoining staterooms and asked, "Ali, are you up?"

"Yes, Shelly, what's up?" she said, slightly frustrated by the intrusion.

"Sorry to interrupt, but I wanted to see if you'd like coffee from the wardroom. I'm heading there for a cup."

"Sure, see if they have some decent breakfast pastry, too."

"Will do. You, okay?"

"Yes, Shelly, I'm fine. I'm a little tired but okay. Why do you ask?"

"You're always so busy. I worry about you."

"Thanks, don't worry about it; it's inappropriate for you to be concerned about me, anyway. After all, I'm the squadron XO, your superior officer. It looks bad, especially when other officers and crew might notice. You should know that by now. Thanks for your concern, though."

Michelle gave a long, frustrated sigh and said: "I suppose so. I'll be back in a few." Michelle never liked the military crap. At best, she tolerated it. At worst, she resented it outright without saying so. She thought people should be on friendlier terms and never subscribed to all the regulations about seniority and rank much.

Ali still didn't fully recognize Michelle's more than friendly interest in her. She felt like Michelle was way too concerned. They were, after all, friends, sure, more like sisters, but she felt like she had to be cautious about how it looked to the crew, her superiors, and the other officers.

She couldn't let it seem like she showed any favoritism or attention to Michelle, even though Michelle was still a little screwed up and often needed her 'guidance.' Ali had a lot of other people to consider, and their opinions of her and Michelle's relationship were an issue. Ali didn't understand how deep Michelle's feelings for her were.

Chapter 2: The Operators

The Seal Team's briefing

A group of Special Operators, in this case, a team of Navy seals, were to be dropped at a target area near Al-Tanf. They'd been assigned to track and engage a group of ISIS rebels. A very special target: Yousef Al-Sherif, a high-level ISIS lieutenant who was close to none other than Abu Musab al-Zarqawi himself, founder of ISIS and current leader of the 'ISIL.'

Master Chief Brooks, the OIC (officer in charge) of Seal Team 'Charlie,' was an eighteen-year Navy veteran with a long list of combat assignments and action in Iraq and Afghanistan. He and his team were heading into the de-confliction zone in Syria, around the Al-Tanf compound, for the first time. Gilbert Greene, Faustino Garcia, Max Muldoon, Greg Mallory, Tony Calabrese, and Shane Sandoval made up his team—all seasoned operators.

Greene was the Chief's second. Sandoval was another Texan from the same hometown Ali was from. They weren't acquainted, but Ali knew of him. She was an aviation officer and squadron XO, a reasonably senior one. Sandoval was an enlisted man and a Seal team member. They rarely socialize or 'fraternize,' Navy Regs again. It's a class system.

The rest of the Seal team were from all over the country. Calabrese, a lanky and gregarious New Yorker, and his bravado showed it. He had a knack for saying too much at the wrong time and pissing people off, Like Michelle did.

You could tell by looking at him. He had that "New York state of mind" type of demeanor. Greene, Garcia, and Mallory were all from California: Greene from the L.A. area, Garcia from the Central

Valley, and Mallory from Julian, a small mountain town in North San Diego County. All were veterans of Iraq and Afghanistan.

Before flying to the op area around Al-Tanf, Brook's team gathered for their pre-mission intel briefing. Brooks stood in the front of the brightly lit briefing compartment surrounded by his team: Petty Officers Greene, Garcia, Muldoon, Mallory, Calabrese, and Sandoval—all with years of field experience. The mission ahead was dangerous, but these guys were ready for whatever came their way.

Brooks piped up. "Okay, listen up. There's an intel briefing in five minutes. The ship's intel officer will outline mission parameters and objectives." A few minutes later, the CVIC lit up, and Stefanik appeared. He opened his briefing with a description of the mission.

"Good day, gents. This briefing will provide you with the details for today's op. The mission will be in the de-confliction zone around Al-Tanf in Eastern Syria, a high-risk, high-stakes operation. Your target is Yousef Al-Sherif, a high-level ISIS lieutenant very close to Al-Zarqawi himself."

"As most of you are probably already aware, ISIS overran the garrison at Al-Tanf last year in May. Since that time, Syrian rebels, backed up by our coalition partners and our guys, recaptured it in March. Satellite imagery, as well as ground intel, indicates a group of ISIS fighters forming up on the Iraqi side of the border near it."

"Ground intel confirms their target is the garrison at Al-Tanf. They're likely planning on trying to retake it. We know at least 50 of them are in this group, maybe more, with Al-Sherif leading them, but it's unclear what their next move is or when."

"Your objective will be to capture Al-Sherif alive or eliminate the threat! You will infiltrate their camp, wherever it is located, either in Syria or Iraq, wherever you find them, or whatever the case may be, gather as much intel as possible, and capture or kill Al-Sherif before they can mount another attack. You'll need to be in and out quickly and quietly."

Garcia wanted more and couldn't resist asking, "Any specific details on their location, sir?"

Stefanik's response was direct, droning, and robotic sounding. "They're being tracked by satellite and Predator imaging. Their location will be relayed to Master Chief Brooks and your team before you arrive at the insertion point. There will be three helicopters in Specter flight, with Lieutenant-Commander Cabrillo in command of the lead bird."

Garcia thought his droning briefing was odd: *"¡Por favor, no te metas conmigo, hombre!* Stefanik must still be asleep. What a boring briefer he is." He kept it to himself, though. Garcia didn't think Brooks would appreciate his sarcasm, anyway. Garcia went on with a second question. "What's our extraction plan, sir?"

"Lieutenant-Commander Cabrillo and her team will drop you off at a location determined by the target's location and pick you up at the nearest extraction point. That depends on where you contact them. Master Chief Brooks has a list of all extraction points. Do you have questions, gents?"

Brooks added his part to the briefing for his guys, "No, Commander, that covers it, I think. We'll contact the extraction team with our location after we acquire the target and whatever intel we can collect and have Al-Sharif."

"What about communication, Master Chief?"

"We'll use encrypted radios and sat comm as a backup. Silence is golden, guys. 'NO' radio traffic until we're clear of the target."

"Any specific gear we need to bring, Master Chief?"

"Standard load-out; night vision, silenced weapons, and extra ammo. Be prepared for anything."

"When do we leave?"

"Muster at 03:30 tomorrow at the debarkation station. We depart at 04:00, so be ready and stay sharp. I don't need to remind you that success depends on each of us, do I?"

"No, Master Chief!"

Stefanik concluded, "Very well then, gents. Master Chief, take charge." That was it; Stefanik broke off the meeting. Brooks continued after the viewer goes dark: "Okay, prep your gear. Make sure about everything and try to get some rest. You'll need it!"

Brooks seemed more fidgety and on edge than usual. His brown eyes darted left to right, looking everything over and scowling as he walked around the briefing room. Greene, a first class and the squad's communications specialist, noted Brooks' more-than-usual OCD-like behavior, so he did what he was supposed to do and asked,

"Master Chief, are you okay, man? You seem really on edge for this one. Anything bothering you?"

"No, no more than usual. Don't sweat it, Gilbert. I'll be all right! Make sure the com equipment is working, and I mean well, and bring spare batteries.! If it fails, we're fucked for extraction! I'll be okay. I'm more worried about all of you than I am about myself right now."

Brooks always played down his own OCD tendencies and hypervigilance as a concern for his guys. Greene thought it was much more than it usually was for him this time.

Garcia, the corpsman, had been checking over his gear when Greene walked over to him. He spoke in a low tone so Brooks wouldn't notice or hear what he was saying.

"Tino, does it seem like Master Chief's more on edge to you than usual?"

Garcia didn't want to get into it, but he agreed. "Yeah, a bit more than usual, but you know him, Gilbert. He's always amped up before we go out on an op."

"True, but he seems more worked up about something than usual. Just watch him."

"Sure, will do Gilbert."

Brooks and His Daughter Angie

Garcia was no doctor, but he was a very experienced Navy combat corpsman who recognized the symptoms of traumatic stress when he saw them. Brooks had them in spades. He'd been going to Navy shrinks about it for some time, but they always cleared him for duty. Seal team leaders were, after all, valuable assets that are difficult to come by.

Brooks thought it was all bullshit on the outside, but on the inside, he knew he wasn't all there and wasn't the man he'd once been. His wife had divorced him a couple of years earlier and alienated his son so much that the kid barely spoke to him anymore.

Brook's daughter Angela was a college coed at San Diego State. She'd realized what was going on with him and did her best to spend as much time as possible with him, trying to make sure he got help the best she could. No one loves you like Daddy's little girl!

She put up with a lot of his crap for it, too. She understood why he was like he was, so she just let it slide and kept trying to help her dad. Her conversations with him were always the same. "Daddy, are you okay? Have you been going to your appointments with the Navy doctors?"

"I'm okay, Angie, I'm okay."

"Oh, you're so not okay, Daddy! Whenever I'm here with you, I hear you toss, turn, and talk in your sleep! You pace around the house in the middle of the night! You're distant from me and withdrawn from everything that isn't work-related."

"Please go to therapy and let them help you." She grabbed him by his forearm, pleading, "I'm begging you, Daddy, please! Go to your appointments!"

Brooks would always respond the same way, "Okay, princess, I promise I'll keep up with the meetings and take their damned pills! I'll do it for you, okay?"

"Just try. That's all I'm asking."

"Okay, Angie, okay!" was his usual answer. She knew he probably wouldn't, though. She kept in touch with his navy psychologists and tracked his attendance at therapy sessions. Angela Books wouldn't give up on her dad, even though it seemed everyone else had!

Truth be told, Brooks was lucky to have a daughter like her. He didn't have anyone else who cared about him except his guys. Seals look after each other. Muldoon and Garcia had overheard Greene and Garcia's conversation. They just looked at each other and shot a facial expression like "WTF, seriously" to each other without even saying a word.

They'd both heard what was said. Both were smart enough to keep their mouths shut and just went on checking over their gear. Making sure weapons were ready and in top condition was more critical than bullshit and opinions to them. Calabrese, the explosives specialist, was too preoccupied with his area of expertise—prepping breaching charges used for clearing out and blowing up 'obstacles'—even to listen.

While all this was happening, Sandoval, the team's sniper, studied areas mapped out around the fort's perimeter that were identified as potential threats. He tried to pick up some valuable intel they might have overlooked, searching for good shooting positions he could use.

The camaraderie and screwing around grew more measured, serious, and quieter as the op drew near. They knew they could count on each other, so they looked past petty concerns or disagreements. In a fight, they knew they had to. Their lives depended on it. But that didn't remove the anxiety.

POG's; "Private Contractors"

The "contractors" never told anyone anything other than their first names—probably not their real ones, either. These guys want to stay as anonymous as possible for operational security and their own and their families' security and safety. None of them ever carried identification with them on missions.

The Seals considered these guys 'POGS—Personnel Other than Government, or grunts to some.' They never actually spoke or interacted with anyone much unless they had to. Brooks and his guys weren't sure if they were CIA or private contractors. They all wore desert camo BDUs and keffiyeh scarves around their necks, the bandanas one always sees in photos of ISIS or Al Qaida, and scruffy, untrimmed long beards to blend in easier with locals.

These guys carried nonstandard weapons. Czech VZ 58 rifles, AK 47s, or some variation. They looked like they were ISIS fighters or local Muslim militia types themselves. They were never very chatty. When they did, it was easy to tell they were contract spooks.

Looking them over closely with his usual critical eye, Brooks surmised they were probably infiltrating either into Iraq or into ISIS territory in Syria. It was unclear what they were up to, but he wasn't supposed to know and wouldn't ask. Whatever it was or wherever they were going, it was sure to be hot.

Seal teams had a lot in common with these guys. Still, they were Non-military private contractors or CIA. More likely, private security assassins hired to eliminate somebody American troops couldn't legally kill. Probably hired by somebody who wanted to be sure that the U.S. government couldn't be tied to the deed "Plausible Deniability."

Brooks didn't rule out the possibility of murder for hire, either. Some private vendetta their "sponsor" hired them for. He just told himself, "That's how it is. It's their karma to deal with. Time to mount up and get ready."

Marines

A group of marines from the MEU 'Marine Amphibious Unit' also embarked on the ship and were going on the flight with SEALS. These guys were the grunts, the diggers, the principal fighting force working out of the garrison compound. They trained the local SDF forces and only joined in a fight if necessary, but they would respond with a potent defensive effort if attacked.

These guys got the worst assignments, 'Shit jobs,' since there were no civilian laborers. Their days comprised training with the local militia, working out, cleaning and maintaining weapons, and guard duty.

They frequently patrolled the de-confliction zone outside the compound with the local forces and whatever else was assigned to them. The standard description for all services is "Hurry up and wait." In this case, waiting for something to happen.

The pre-mission briefing took place in the platoon spaces on the ship. Since they were Marines, their platoon commander, 1st Lieutenant Summers, gave the mission briefing:

"Okay, listen up! Your objectives during this assignment are One: relieve the alpha squad working out of the garrison. Two: support training efforts for the SDF operating in and around the de-confliction zone, and Three: other duties as assigned. Your presence at this base is vital and strategic in this ongoing conflict."

"It's a hot combat zone, so listen to Staff Sergeant Reynolds and pay attention! Staff Sergeant Reynold, take charge of the squad."

"Yes sir, Company Attention!" the Lieutenant departed the squad bay. "At ease. Gear up! Make sure you've got everything packed and ready," Reynolds added his directions to the briefing.

In the Corps, a squad leader, typically a sergeant, leads a squad of 13 marines: the squad leader himself and three fireteams of four men in a rifle squad. Staff Sergeant Jake Reynolds headed up this group. His team had been assigned to the Al-Tanf garrison to relieve another group rotating out.

They were a typical bunch of young Marines. Loud, proud, and obnoxious - full of bravado 'piss-and-vinegar,' bragging about how they were going to kill everything and everyone who opposed the USA.

The guys on the Seal team got a kick out of them. When they were together, they knew these guys were green. Fresh out of training and about to get schooled in the realities of war-fighting. Reynolds himself wasn't so eager to get into the shit himself. He knew what was going to happen.

Reynolds barked at his squad, his voice cutting through their 'esprit de corps' bullshit session with each other: "All right, listen up, you hounds. We're going into a pretty hot combat zone. Pay attention and keep your shit wired."

"I don't want to have to write letters to your mommas and tell them what dumb asses you were for getting yourselves killed because you didn't listen to me, so keep it tight, keep it clean, and most of us should be all right. Let's get it done!"

"Oorah!" The Marine cheer went up in unison from his guys. Brooks had met Reynolds a few times on board the ship and saw him put his guys through their paces during physical training on the ship's hangar deck. He saw much of himself in Reynolds. Only other combat vets notice that "stare" and know why once they're blooded and combat-seasoned.

They looked everything over for details and dangers unseen by ordinary men and women. Looking for exits and ways out of trouble, any advantages. They had both done multiple deployments to Iraq and Afghanistan more than once and had seen their share of bloody action. Both would rather forget about it but never could.

This time, it was Syria. For Reynolds, these boot camps were way too eager. He just tried to temper them and inject a little reality and fear into their inexperienced heads to prepare them for what would likely come their way soon.

Chapter 3: ISIS

Background

Yousef Al Sherif, leader of this ISIS cell inside Northwestern Iraq near the border with Syria, had been away visiting his network of other cell leaders, discussing and planning a significant assault on the American-held compound at the border of Iraq and Syria.

The target in question was a large base that had grown out of a border checkpoint during the Iraq war with the Western infidels "as they saw it." After U.S. forces and their coalition allies established it as a forward operating base, they designated a large ring of territory known as the 'de-confliction zone.' The goal of which was to eliminate ISIS forces from the zone.

Its legality was murky as it is sovereign Syrian territory, and the USA had not received an invitation to intervene. Since ISIS had staked its claim on the entire Levant, including Syria, the battle lines were drawn.

Yousef and his ISIS fighters planned and executed many raids on the entire area in the past. Today, he met with other leaders and field commanders in his territory to discuss and plan his next move: an attack inside the Al-Tanf de-confliction zone on the base itself.

The Shura Council

The meeting with the military council began with the usual formal greetings:

'As-salām 'alaykum' brothers "عليكم" , *Peace be upon you*. A chorus of responses arose from the group 'wa alaykum as-salām' "مالسيكمعل" *and peace be upon you*" brother.

Yousef began to speak and make his pitch for support.

أيها إلخوة، لكن ان ؤ من بر وي ال خ ل ف اق بر اه ي ف ل ي ت ع ادة خ ال ف ة الت ف ي ق دن اه ال ى أي
الفار الابو ينطاي ي ن و الس ي ي ر ي ق ع ل ى ت ظ ن ا آل ن و ا جت ب ص ح ح ٔالخ طاء و إنش اء
أمة اس ال م ي ة ق دس ة و عدالت ة ع ي ش ر ب ك الش ري ع ة الإس لام ي ق و ال ع دل إنش اء الله، المت ج م ي ع
ال م س ل م ين ! ئ ج ب ع ل ي ن ا ل ه س و اي ك س ي ي ك ل ف ت ق اس ي م غ ي ل ي ق ان ون ي ل ارل ض ي ل خ ال ف ة
إلف م ر وض ع ل ي ن ا، مرة واحدة وإل ى أل ب د"

"*Brothers, we all believe in the vision of Caliph Ibrahim. In the caliphate's restoration, of that which was lost to us at the hands of the Infidel, British and French. It now falls to us and is our duty to correct these wrongs and create a holy and just Islamic nation that lives by Islamic law and justice, inshallah, a nation for all Muslims! We will end Sykes-Picot, the illegal division of caliphate territories forced upon us, once and for all!*"

A murmured agreement went up from the group.   Yousef continued:

أنا أق ف أم ام ك م من ال ي وم، أم ام ال ج ل س من اق ش ة خ ظ ن ا لاست ع ادة ق اع ل ف ك ف ارف ال ي ت ن ف
الت ي ف ق دها مؤخرا أم ام ال ج ي ش ال س ور ي ل ح ر و الأم ي ي ي ت ج ح ال ف م م ض ن ا ج ن ود ي
رج ال ش ج ع ان ف ي ع ل ون ما موض ر ور ي ل خ وض ه ذال م ع رك ة ت ق ت ل الك ب ر عدد مم ك ن
من ال غ ز اة ل ف ار ل ل ك ن اي ن ح ت اج ال ى م س اع دت ك م الأس ل ح ة ق و ال ن ادق ال ذ خ اي ر و ال ص ور اي خ
وأي إم داد الض اف ي ق ي م ك ن ال ح ص و ل ع ل ي ه ا.

"*I Come before you here today, before the council, to discuss our plan to recapture the Infidel base at Al-Tanf that was lost recently to the SDF, the vile Americans, and their coalition of infidels against us. My soldiers are brave men and will do what is necessary to fight this battle and kill as many infidel invaders as possible. Still, we need your help with weapons, rifles, ammunition, rockets, and additional supplies.*"

One of the senior members of the Council meeting responded:

أل خي يو س ف، هذه خطوة جريئة. ماذا تعتقد ون أن الأمير كيي ين وال فصائل ال سورية
سي ف عل ون لف قا ما ف إذا اف ق نا عل خط ت كم؟ و من المر جح أن يعتر ض مجلس ال شورى
على مثل هذه ال خطو ة الجريئة أيضا. أنت جندي شجاع للدولة ال إسلامية ولكن هذا
الهجوم خطير لل غاية. ن ي جب أن ت ض ع في ع ت با ر ك ال ث ي ر ال ذي

"Brother Yousef, this is a bold move. What do you think the Americans and the Syrian factions will do in retaliation against us if we approve your plan? The Shura Council will object to such a bold move as well. You are a courageous soldier for the Islamic State, but this attack is hazardous for us. You must be mindful of the impact you will have."

إن ني أت ف هم مخاو ف كي ا أ خي و لكن إذا تر د د نا بس ب ب ال مخاط ر ال مت صورة ف إ ن قض ت نا ق د
ف ق د ت قبل ال ف ع ل. لا يم كن نا إخراجهم من أرض ال مسلمين إذا كنا خائفين من قتال أعدائنا
أو خائفين ي ما سي ف ع ل ون ه ل ل ان ت قا م ا

"I understand your concerns, brothers, but if we hesitate because of perceived risks, our cause is lost before we begin. We can't drive them out of Muslim lands if we are afraid to fight our enemies or are fearful of what they will do in retaliation."

Yousef had made an interesting argument to the members of the council. They decided it was time to take a vote on the matter up or down.

أ ي رجى ال ا ن ت ظا ر ف ي ال خا رج ي ن جي ن ف إ ن نا قش ط ل ب ك و ن ق ر ر ما إذا كان يمكن نا دعمه أم لا
و ل و ص ي ب ه ق ب ل مجلس ال شورى لل مو اف ق ة عل ي ه أي ض ا

"Brother Yousef, please wait outside while we discuss your request, decide whether we can support it, and recommend it to the Shura Council for their approval."

He left the room where they had just met, satisfied with himself for making a compelling argument, and decided he would have some tea while waiting for their decision. He'd traveled here to Al-Raqqah precisely for this meeting.

Yousef waited for their decision in the main reception hall of this once ornate building, now scarred by bullet holes and broken windows from the combat that had wrestled control of the city from government forces, now the center of this area of ISIS-controlled Syrian territory.

Yousef couldn't help but reflect on his life and how it had taken him to this point, from those terrible days in Ramadi during the war with the infidel Americans. Now, it was a war to liberate Syria and the Levant for Islam.

Ramadi 2006

Yousef was Iraqi by birth. His family lived in the city of Ramadi, in the Al Anbar province, in the latter half of the 1980s. During the U.S. Invasion of Iraq, 'OIF - Operation Iraqi Freedom,' he was already a teenager. He grew up fast and hard during those dangerous Iraq war years, seeing his father and two brothers killed during the house-to-house and street-by-street combat with U.S. and coalition forces during raids on the city.

He, his father, and his brothers had been lucky earlier and had been spared during the bombings, but he'd seen many friends and relatives killed, along with American Soldiers and Marines.

Those experiences as a teenage boy and then as a young man hardened him to the fundamentalist Islamic cause. He felt he must avenge his family and help create a kind of nation and world that they had fought and died fighting for; "An Islamic nation for all Muslim people." It began for him during the last days of September that year.

Yousef's heart pounded as he crouched in the ruins of what had once been their family home in Ramadi. The echoes of gunfire reverberated through the streets, mixed with the distant wails of the wounded and the constant roar of American military vehicles. September 2006 was a time of blood and sorrow for him and countless other Iraqis like him.

He remembered the day vividly when his world had been shattered. It started like any other day, oppressive desert heat pushing down on the city like a suffocating blanket. Then, the Americans launched their relentless assault, pounding the city with artillery and raining death from above.

His father and brothers had been caught up in the crossfire, planting IEDs in the roadway while others were firing mortars and RPGs at the Americans advancing on their location. Their lives had been snuffed out in an instant by the indiscriminate fury of the war. He watched in stark terror as their bodies were torn to pieces by rifle fire and M203 grenade explosions from the advancing Americans.

He was lying motionless amidst the rubble and blood-stained and dusty ground that used to be their neighborhood.

The anguish and pain that tore through him that day left scars that would never heal. Amidst the wreckage of his shattered neighborhood, a fire ignited within his heart and soul, fueled by a mix of grief, rage, pain, and a terrible thirst for revenge. He swore to himself that he would make those responsible for his family's murder pay dearly for their crimes.

So, when the call to arms came from the shadowy figures of ISIS years later, he answered without hesitation. Joining the ranks of ISIS, Yousef found a genuine camaraderie with fellow fighters who shared his pain and thirst for vengeance. Under the black ISIS flag, he learned to wield weapons with deadly proficiency, skills honed by the crucible of conflict that Iraq had become.

Even as he fought alongside his newfound ISIS brothers, the ghosts of his past haunted him relentlessly. Night after night, he was plagued by nightmares of those days in Ramadi. The faces of his father and brothers twisted in agony as they were ripped from him in a storm of violence.

The lines between friend and foe blurred in the chaos as Yousef questioned his cause's righteousness. But the memories of his father's voice, his brothers' laughter, and the searing pain of their loss drove him forward, propelling him deeper into the blackness of that conflict.

In the chaos of Ramadi, amidst the clash of ideologies and clash of arms, Yousef's story was just one among many. His was a tragic tale of the ravages of war and the unyielding grip of vengeance that threatened to consume everything and everyone.

A Desert Vision

Yousef often reminisced about times in Ramadi before the war with the Americans. He cherished the memories of his father, brothers, mother, and the many pleasant days they had as a family. He missed those times, resenting and hating those who took them. It made him an easy convert to the ISIS cause.

The desert in Syria stretches out as far as the eye can see, dotted with hills surrounding the Al-Tanf garrison compound. As nightfall took hold, the temperature dropped. A warm desert breeze still kissed his face and caressed his bare arms. The sky became a breathtaking display of orange, pink, and purple hues, blending into a stunning display of colors at dusk.

That evening before the raid on the garrison at Al Tanf, he sat alone in the desert, relaxing in the evening air and drinking his unique blend of Turkish-style coffee. Then, he performed the fourth Maghrib prayer and meditated on what was to come, considering its outcome and the brothers who would be martyred.

"ستنجح هذه الخطة وسندمر هؤلاء الكفار بإذن الله".

"This plan will be successful! We will destroy these infidels, God willing."

He mulled over the plan's finer points, sitting alone, trying to think of anything and everything he might have overlooked. He was stressed, nervous, and anxious, trying to be sure he'd thought of everything.

A Western psychiatrist would have diagnosed him with symptoms of PTSD himself: obsessive-compulsive tendencies, hypervigilance, fits of extreme anger, and often trouble sleeping. He could never forget the blood and carnage he'd endured in Ramadi as a youth. Now, as a man, he was facing bloody combat here in Syria, fighting infidels, Assad's government forces, Syrian Democratic Forces (SDF), and other factions of apostate Muslims.

The evening light faded, and darkness enveloped the desert. Only the glow of stars and the moon filled the sky and illuminated the landscape. The silence almost seemed tangible, broken only by occasional wisps of wind rustling through the dunes. A scene of peaceful serenity invoked an unexpected and strange state of mind.

Tired from the day's preparations, sitting alone in the desert, fatigue overcame Yousef, and he drifted off into a dream-like meditative state, a reverie, experiencing lucid images and strange thoughts. As it took shape in his mind, a Djinn appeared to him like a smoky apparition, paying a surprising and frightful visit.

In this state, the Djinn spoke to him in barely audible whispers and surreal tones, taunting him, snickering, and saying:

يوسف الشريف، هل تعتقد أنك ستنجح في هذا الجهاد ضد هؤلاء الكفار؟ ستفشل وستدفع ثمناً باهظاً لغطرستك إذا كنت تعتقد أن هذا ما يريده الله منك ومن إخوانك ستموتون جميعاً ما لم تطلبوا مساعدتي وتقبلوها

"Yousef Al-Sherif, do you believe you will succeed in this Jihad against these infidels? You will fail and pay a heavy price for your arrogance if you believe this is what Allah wants from you and your brothers. You will all die unless you ask for and accept my help!"

The Djinn continued its sardonic taunting of him.

يجب عليك أن تعبدني سأجعل معركتك ناجحة

"You should worship me! I would make your battle successful!"

Yousef could barely hear its laughter, like a whisper in a dream, but still audible in his mind, as if he were an imbecile who just wanted to tease him. He returned to reality as the taunts and jeers trailed into silence. Incredulous and dumbfounded at what he'd seen and heard, he wasn't sure what to make of it all or how true it would be.

He'd known men from Egypt who had used a drug called 'The Blue Elephant,' a derivative of DMT, a highly psychoactive hallucinogenic drug. Those who used it experienced encounters with Djinn, but he'd never heard of anyone seeing them in a mere half-awake, half-asleep lucid dream state. Was it a dream, a vision, or was he hallucinating from the day's heat and exhaustion? Maybe it was just his fear and doubt!

Nassir was passing by where Yousef had been praying and contemplating the battle to come. He noticed Yousef's doubtful look of fear and surprise as he recovered from this strange reverie.

He asked Yousef:

"أخي، هل أنت بخير؟ يبدو أنك مشوش الذهن. ما ألمر؟"

"Brother, are you alright? You seem disoriented. What is the matter?"

Yousef told him the story of his vision, the Djinn, of its taunts, and how it spoke of his group of fighter's failure to prevail in the battle. Nassir replied,

يا أخي، إن الله تبارك وتعالى لن يرضى عن الم هذاالجني الشرير فين وصراعنا مع هؤلاء الكفار ثم يقترح عليك أن تعبده بدلا من هكذا!

"Brother, Surely Allah, blessed be he, would disapprove of this evil Djinn's talk against us and our struggle with these Infidels, then blasphemously suggest you worship him instead!"

Yousef had to agree with him.

"نعم أخي لقد كانت المية وعي نقدر يرظل خاعنا وظناع يطل الخلي عن هجومنا على هذا المكان."

"Yes, brother, it was surely an evil spell to trick us and persuade me to give up our attack on this place."

Deep down, creeping doubt rose about what the Djinn had suggested. Nassir asked him,

"كيف كان هذا المكان يا أخي؟ ماذا قال لك الجن؟"

"What did this place seem to be like, brother? What did the Djinn say to you?"

Yousef looked at his friend with a fearful expression and replied,

"كما قلت من قبل، عندما تحدث إلي أخبرني أنه يجب علي أن أعبده كما لو كان إلها وفي المقابل سيمنحنا المعركة! كانت الرؤية أشبه ببرزخ ذلك المكان بين هذا العالم والسماء"

"As I said before, when he spoke to me, he told me I should worship him as if he were God, and in return, he would give us the battle! The vision of it was like Barzakh — that place between this world and heaven."

"لقد تمكنت من رؤية مكانين من بعيد على جانبي بعيدًا جدًا قد كانت صورة غريبة
ومخيفة يا أخي، ومع ذلك لم أكن نائمًا بل كنت أرتاح وأفكر في مسار عملنا!"

"I could see both places far off in the distance on either side of me, far away. It was a strange and frightful image, brother, yet I wasn't sleeping, just resting and contemplating our course of action!"

"عندما تحدث بدا ألمًا وكان صوتًا واحدًا ولكن أصواتًا عديدة تردد ما قاله
في نفس الوقت بدا ألمًا وكان الصوت نسيمًا ناعمًا على وجهي! همسة من إتجاهات عديدة"

"When it spoke, it was as if it were a whispered voice, but many voices echoing what it said simultaneously. It seemed as if the voice was a soft breeze on my face! A whisper from many directions!"

Nassir was dumbfounded at his description and asked,

"لابد أنه يكون هذا من قِبَل جنٍ شرير للغاية ليقترح مثل هذا ألمًا! إنه أعظم من الله
قلس تبارك تعالى"

"It must have been a very evil Djinn to suggest such a thing! That it was greater than Allah himself, blessed be he!"

Yousef no longer felt confident about his plan. He struggled to forget the strange vision and concentrate on the tasks at hand, and it nagged at him. It made him feel like he should see what kind of frame of mind the rest of the men were in before the fight, so he walked about talking to them, looking at their faces and into their eyes.

Most were quiet. He could see the fear in their eyes. There was stark terror in the younger, less experienced men. Most were resigned to their fate, whatever happened. All had taken their last rites. Walking around and seeing their fear, he felt it was time to address them and give them a pep talk, thinking it might shore up his rising doubts and fear. He stood in front of them and spoke.

"أيها إلخوة فتقوّوا وقولوا بأنكم ربما تكون هذه المعركة هي الأخيرة تشيرين من اولكن
تأكدوا أن بلدن الملك افأةالموعود قد نكن الجنة مع الله باركوت على الأبد! إذا
كان منكم فسوف نتصر ونطرد الكفار والمرتدين وأولئك الذين ارتدوا عن الإسلام
منا! هذه الأرض أرض إسلامية. وسوف نحررها من كل هؤلاء"

47

"Brothers, be of strong heart. This battle will probably be the last for many of us. But be certain we will all receive the promised reward and dwell in heaven with Allah, blessed be he, forever! If he is with us, we will prevail and drive the infidels and apostates, the Murtadeen, and those who have turned away from Islam from among us. This land is Muslim land. We will free it from them all!"

At the end of his speech, a chorus of "Allahu-Akbar, Allahu-akbar" arose among them. Yousef felt his speech had made a difference and quelled their fear and apprehension of the coming fight. But deep down, he wasn't so sure.

It was soon time to move, take up positions, and execute the battle plan. He and his fighters crept under cover of darkness and holed up in the hills near the infidel compound at Al-Tanf on the Iraqi side of the border. It was on the Iraqi side of the border, near the checkpoint. They positioned themselves close enough to execute the plan quickly but far enough away not to attract unwanted attention or be easily detected.

An Unexpected Advantage

Yousef and his men started approaching the compound, setting up their assault points. As they moved in on the compound, they heard the approaching rumble of Specter flight. From experience, they all knew what type of helicopter these were by their distinct sound. Seahawks and their counterparts, Blackhawks, have a distinctive whirring, rumbling, snarling sound, like a swarm of giant angry murder hornets inside a spinning tornado.

The rotors beat and swirl as the wind from these machines rushes down in a cyclone, descending, and the roar of these machines builds to a crescendo as they approach. The noise rises to a deafening level as they approach from above, whipping up whirling clouds of dust and flying debris that churn about like shrapnel in a bomb blast.

Specter flight flew in low over the desert. Yousef realized that knocking them down would be an excellent diversion. In his mind, shooting them out of the sky made his plan even better and more likely to succeed. He felt that familiar adrenalin rush coming on and gave the order to Nassir.

"أسقط وهم لدينا هذه الفرصة الممتازة حضيرة يلزلها جوم على الحصن وتحويل الانتباه عن أبراج الحراسة سيدخل السيارة الملغومة بعد ذلك ثم إطلاق الصواريخ على الأبوابة الأمامية في هذا جنين"

"Take them down! That will give us an excellent diversion to set up the attack on the fort and divert attention from the guard towers. The car bomb will go in next, then the rocket fire on the front gate! This might even work!"

Chapter 4: Un-Expected Action

Zero Dark 30

The Al-Tanf mission was scheduled for very early that morning. It was in the border area between Syria and Iraq. The mission was to transport a Seal team to relieve another team rotating back to the United States.

They were also carrying a Marine rifle squad, a group of non-military Special Operators hitching a ride with them, a bunch of supplies, and a package of classified equipment to the base.

The scuttlebutt (rumor) was that these Special Operators were CIA or some kind of outside contractors, there for who knows what tasking. No one's ever sure what these guys are up to when they show up but the higher-ups.

Ali flew the lead bird. Before the briefing, she and Michelle reviewed her detailed mission plan together. It called for three helicopters, routing vectors, and alternative routes.

They reviewed weather predictions and potential issues or threats that might arise together—all the little details to consider when planning a mission like this. Ali wrapped it up with, "Okay, that's it. We'll go with what we've got for now. There'll likely be changes after the Intel briefing, anyway."

"Dang Ali, this seems pretty thorough to me. Do you think it will change?"

"I'm sure of it; Stefanik always changes something."

Ali added jokingly, "That's it, then. Let's head for the ready room, Deadeye."

Michelle laughed and said, "Aye-Aye XO!"

Michelle's handle, or 'call sign,' was Deadeye. She'd got it during flight training. As a teenager, she'd been an avid target shooter and accomplished marksman with perfect eyesight. She took great pride in her precise shots and excellent aim. Her civilian friends and Navy buddies just started calling her 'Deadeye.' The name just stuck.

They both went to the squadron's ready room for the ship's intel officer's briefing. Ali was proud of being a meticulous planner. She strove to be as thorough and ahead of the curve as possible. She always had a detailed mission plan ready before Stefanik's briefings, but he invariably changed something during briefings.

The MH-60 is a special-purpose, multi-mission helicopter used for clandestine missions, ASW ops, transporting troops and equipment, and many other missions. It's a general-purpose machine but with some special "features."

Mission Briefing

The two of them and the other squadron officers assembled in their ready room for the briefing over the ship's CVIC, or 'Combat Visual Information Center,' a closed-circuit secure video system. It lit up, and the intel briefer, Commander Avril B. Stefanik, the briefing officer, appeared.

Stefanik opened his briefing by describing the mission: "Good morning, ladies and gents; your mission assignment will be to transport the embarked seal team, a group of marines, and other non-military personnel to a FOB 'Forward Operating Base' near the Iraqi border."

"You'll be transporting replenishment supplies and a special equipment package. Your flight will comprise Lieutenant-Commander Cabrillo's aircraft as lead ship, Lieutenants Lewis and Spaatz, and their respective crews, passengers, and gear as previously outlined, flying the number 2 and number 3 ships. Your destination is the compound at Al-Tanf."

Here, the term 'Special equipment' wasn't specific. It was all classified gear for use on the base by the Seals and the POGs working out of the base. The term POG—Personnel other than Government, or Grunt—is dubious, usually meaning CIA, outside contractors doing dirty work for them, or some other three-letter agency.

The destination Stefanik was briefing on raised a few eyebrows, including Ali and Michelle's. Its location was some pretty hotly contested real estate that had changed hands more than once. Stefanik continued,

"As it's a classified mission, it will be a night launch and an early morning flight. Fly in, drop off, unload, then fly out as quickly as possible." All-in-all, a potential shitstorm.

At this time, in Eastern Syria, this compound was in ISIS territory. Russians, Americans, the SDF - 'Syrian Democratic Forces' rebels, Kurds, and Syrian government forces battled back and forth with one another for control of it, along with several other local militia groups and factions. It had become a take-loose and retake bloody dance of combat, with many defenseless civilians caught up in the crossfire and malaise.

Ali and Michelle exchanged quick, frustrated glances, concerned about the potential risks. Michelle's grimaced and frowning expression was more blatant and caught Stefanik's eye. He piped up in a terse and demanding tone. "You seem concerned, Lieutenant Robbins. Do you have concerns you'd like to share with us?"

"No, sir, I have nothing to add," was her sheepish, embarrassed answer at being tagged. Stefanik noted that, too.

"Perhaps you should discuss your concerns further with Lieutenant-Commander Cabrillo then."

"Yes, sir, I'll do that." She felt stupid for being singled out like that, showing what Stefanik considered "undue concern" so blatantly and told herself, "Shit! I'll never let that happen again."

Stefanik was a little aggravated by her reaction. He interpreted it as her displeasure and disagreement with his briefing and mission assessment.

Ali turned her head, glaring a little at her co-pilot, but wasn't that surprised, knowing Michelle as she did. Privately, she thought it was a little funny. Michelle was always overly opinionated and emotional; some might have called her 'Obnoxious and mouthy.'

The terms 'bearing' and 'decorum' often get thrown around. Officers and enlisted people aren't supposed to express or display negative feelings or comments to or about senior officers. They're expected to convey support and 'proper decorum.'

Call it 'politics.' Everyone knows and accepts the risks. Questioning or disagreeing with superiors is just interpreted as weakness or contempt, even insubordination, or just being a 'little bitch.' Stefanik concluded his briefing with:

"Very well then, if there are no further questions, the op is scheduled to depart at 0400. Plan accordingly. This completes the

Intel portion of your mission briefing. Lieutenant-Commander Cabrillo, take charge of your squadron."

"Yes, Sir." The CVIC went blank, and Ali concluded with:

"If anyone has questions, let's review them before we adjourn."

One of the other pilots asked matter-of-factly, "Are we sure about this route, Lieutenant-Commander? This is a pretty dicey place to be flying across Syria, too." The question made Michelle feel better about her reaction to Stefanik's briefing. It confirmed to her the concerns she'd had.

Ali's response was predictable, "Well, I'm open to suggestions. If you have anything better, let's hear it!"

"I didn't have any better routes in mind, Ma'am. Just voicing my concern for a risky mission."

"I see. I'm sure you and Lieutenant Robbins share the same concern, but that's the mission. It is what it is! Anyone else?" After that brief exchange of doubts and fears, Ali wrapped it up. "Okay then, dismissed! Lieutenant Robbins, a word, please."

The ready room cleared out except for the two of them. Michelle was sure she was in for a royal ass-chewing for the episode with Stefanik: "Oh shit, here it comes!"

Ali barked at her sternly, like you would scold a petulant child for something stupid. "Damn, Shelly, what was that about? What were you thinking?"

"I know, I know, Ali, I fucked up. I should have known he would catch it. I'm sorry."

"Sorry won't cut it with him, Lieutenant." She made a point of using Michelle's rank instead of her name.

"You can't screw up like that with Stefanik. He's a prick and a ring-knocker! He's always looking for smart-ass comments or what he thinks is a poor attitude. That's one of the best ways I can think of to sink your career."

"Shit like that will end up in a fitness report in a heartbeat! If he thinks you're not on board with everything and aren't on your A-game, he'll have a chat with Commander Deering, and we all know shit flows downhill, don't we! Then I'll hear about it."

"Sorry, Ali. I'll work on it, XO." Ali's tone changed to something more like the same parent talking to the same kid after she'd corrected the kid:

"Just be mindful of who's who and around when you show your cards, okay? That's enough of that of that shit. Go get some rest. We muster at 03:30!"

Pre-Flight

03:00; the 1MC crackled with "flight quarters, flight quarters. All concerned personnel, man your stations!" throughout the ship. The Air Department, Cabrillo, her officers, their crews, and all the other squadron personnel took stations and prepared for launch.

Ali, the other pilots, and their crews assembled, doing pre-flight checks and preparations to board their machines. Michelle briefed Petty Officer Flanigan, the helicopter's AWO, on what he needed to know about their mission.

While Flannigan was getting briefed, Ali did a final walk-around inspection, reviewing the maintenance and pre-flight checklists to ensure the maintenance crews had left nothing undone or overlooked.

Ali checked to make sure all battens 'devices securing flight controls,' 'remove before flight' tags, streamers, and covers were removed. She eyeballed the tires and wheels and checked the machine again for leaks, such as hydraulic fluid, oil, or fuel.

After she'd finished her walk-around inspection, Ali spotted the squadron maintenance chief, Master Chief Banks. Approaching him, she asked, "Everything squared away, Master Chief? Any problems?"

"No, Ma'am, I checked everything over personally. Everyone's done everything on the maintenance sheet for your bird and the others. They're all four-oh squared away! The maintenance crews have done a thorough job."

Pleased by his report, she answered: "Outstanding Master Chief!"

Banks, a thirty-year Navy veteran aircraft mechanic, knew these helicopters inside out and made sure the maintenance crews did, too. Banks made them do everything by the book.

The Master Chief was on his last tour, his 'Sunset tour,' after working on everything from F-14 Tomcats to A-6 Intruders and now Seahawks. Banks worked his guys hard, and they respected him for his skill and experience. He was the 'grandmaster' of aircraft maintenance to them.

Ali moved on from her conversation with Banks and looked at the water from the port-side aircraft elevator opening, watching the weather and sea state. Michelle spotted her, approached, and asked, "What do you think, Ali? Are we ready to go?"

"Yeah, I was just looking out at the water. I wanted to check out what it looked like. You know, the weather and everything."

"Looks okay to me. What're you thinking?"

"I don't know. It's just a feeling. Sometimes, I get funny feelings about our missions. Never mind, let's go."

"Why? Is something bothering you about this one? I mean, other than, it's our typical potential shit-storm of a mission?"

"I don't know; it's just a feeling. Never mind, let's go."

The 'Blue Shirts' rolled each of the three helicopters onto the port-side aircraft elevator one by one, sent them up to the flight deck, and then moved and spotted them in their assigned positions for launch.

Everything was then set for launch!

Debarkation

All three groups heading to the garrison made their way to debarkation stations, a staging area for troops below the flight deck, near the steel access ladders going to the flight deck. The Seal Team, the Marines, and the POGs mustered up with all their gear and weapons in tow, waiting for the go-ahead to climb the ladders to board their assigned helicopters.

Before heading up to the flight deck, they shared a last bit of camaraderie, joking about the mission, yanking each other's chains, and keeping a positive spin. They knew the risks involved but also knew they were the best of the best.

Brooks gave a nod. They go ahead, and everyone heads up the access ladders, waiting for the 'go' signal to board. When it came, they piled into Cabrillo's helicopter first, the rest following into the second and third helicopters, ready to face whatever fate awaited them.

Brooks was a highly respected Seal team leader with 24 years of service, known for his leadership and intuitive tactical skills. He suffered his particular brand of traumatic stress from so many years of high-intensity combat ops, a failed marriage, and kids who barely knew him. As they prepped for this op, he addressed his boys inside the noisy cabin of the Seahawk with a last pep talk.

Brooks had a way of being direct and matter-of-fact with his guys. Greene, Tino Garcia, Muldoon, Calabrese, and Sandoval listened to his spiel like a sermon.

"Okay, guys; listen up; here's a last brief; objectives have already been outlined well; you all know what they are. Infiltrate Hajji's camp, grab Al Sherif, alive, if possible; otherwise, make sure the fucker is dead, and collect as much intel as possible."

"You all know the drill: in quick and quiet, bag Al Sherif or switch him off, grab hard drives, maps, notebooks, whatever you find, out quick and silent! Avoid contact!"

Muldoon asked, "What about the Camp Master Chief? We just going to skulk away and not do any damage?"

Amused, Brooks told him, "We'll get out of Dodge, then drop a few M433 eggs up their asses. See if that wakes them up!" A big laugh went up in unison from everyone at his B.S.!

Muldoon was disappointed, though. He liked to make a lot of noise.

Brooks continued, "Don't sweat it; just in and out and cause some mayhem when we're out of range. Head for the closest extraction point. Cover each other's asses, and come back in one piece, copy?"

"Copy, Master Chief."

Launch & Departure

Ali, Michelle, Flannigan, and the rest had already strapped in, as the turbines spun at idle. The two of them went through their checklists before spooling up engines for takeoff.

Nights and early mornings on a Navy ship at sea are magical, like nothing a sailor can experience on land. The sky and the sea are pitch black in even the slightest overcast weather, except for the tumblehome of the ship's wake from the bow. The churning water bubbles up behind the ship from the screws as it plows through the water.

Clear nights are even more magical. The sky at sea is ablaze with a million stars, and the moon casts bright white light on the water, shimmering like a dreamscape of luminous silvery wave-tops on the dark water! The scent of salt water has a unique quality, too—like the aroma of fresh linen and springtime air after a gentle rain, all rolled up into a single fragrance.

It isn't easy to describe in words. It has to be experienced to appreciate its unique and pleasant scent. The bioluminescence, 'the glow,' of the minerals in seawater casts a distinctive blue-white glowing aura on top of the churning water, giving it a magical quality all its own!

The three machines spooled up for take-off, engines whining. The signal was given. Gear and cargo had been loaded earlier to speed things up and avoid excessive flight deck movement. Then, the final take-off procedures and power-up. The LSE "Landing signal enlisted" gave the signal for take-off and snapped a sharp salute.

Cabrillo's Seahawk lifted off, departing to port (the left side), and ascended into the darkness. One after the other, they lifted off. Spots six and seven followed in rapid succession, forming up on Cabrillo. Michelle made the radio call: "Flight, Specter 1; Specter flight en route objective."

"Flight, Copy Specter-1. Proceed objective."

"Specter-1, Copy Flight."

All three flew standard formation to the Syrian coast, with Ali and Michelle's helicopter in the lead. The ship wasn't that far offshore, about 50 sea miles, so the flight to the Syrian coast was short. The radio call went out as they crossed the coast: "Specter flight is feet dry."

"Flight, Copy Specter 1; proceed at your discretion."

"Specter-1, Copy Flight, Specter 1 out." That was Cabrillo's only response to the ship. Since this was considered a 'clandestine' mission, there would be no radio communications from then on—an order they would soon have to break.

Take them down!

As Yousef had ordered when they spotted the helicopters coming in, Nassir followed his order and picked their best men to shoot down the infidel helicopters. The weapons that the council in Raqqa had agreed to provide them with at the council meeting were a cache of Russian-made RPGs, a kind of shoulder-fired weapon that launches a rocket equipped with an explosive warhead.

Even better were the "MANPADS." surface-to-air missiles that a single person can fire against aircraft, and best of all, American-made Stinger shoulder-fired missiles taken from the CIA annex in Benghazi several years before.

Nassir picked his best, most accurate shooters: Ahmed Husseini, Mostafa, and Amal. They all grabbed RPGs and drew a bead on the helicopters. Yousef trusted Ali to select their best shooters. They'd grown up together and had lost many friends and family members in the house-to-house urban combat in Ramadi.

Crash of Specter 1

The first Rocket hits Specter 1

Their best shooter, Ahmed, zeroed in on Specter 1 and fired. A rocket crashed into the rear section of Ali's Seahawk near the tail rotor, damaging its gearbox, nearly blowing off its mountings, making it wobble wildly out of balance.

The tail rotor blades spun erratically and splintered into pieces, flying off in all directions from the gearbox. The helicopter began spinning out of control, spiraling towards the ground as it lost altitude, pitch, and yaw control, and then another sudden jolt rocked it.

A second rocket slammed home, hitting behind and above the rear cabin door near the starboard (right) engine and the main rotor gearbox, exploding with a concussive blast. The shock and concussion from it nearly shredded the entire tail section.

What was left of it shuddered violently, sending shock waves through every part of the helicopter, still in one piece. Before she could respond, the lights on the control panel flashed red, and alarms blared.

Ali screamed over the radio, "Mayday! Mayday! Specter-1, I've been hit, losing control! I'm going down!" Her voice was frantic but steady despite the chaos. Fingers danced across the controls as she fought to stabilize the bird.

Michelle looked at her commander with terror in her eyes and fear on her face. "What should I do, Ali? What should I do? The starboard engine's on fire!"

Frustrated by Michelle freezing and terrified herself, she still tries to keep her head, and barked out procedures at Michelle, "Emergency procedures, emergency procedures! Activate fire suppression system!"

Splitting her vision between the controls in front of her and the flames licking the cabin, Michelle calmed down and activated emergency fire extinguishers. The cabin was filling with rising smoke as the fire extinguishers started putting out the flames.

Michelle shrieked, "Fire's under control; we're losing altitude fast!" Flickering emergency lights illuminated her face.

"Fire out, losing altitude, Roger that," Ali yelled back at her, hands gripping the cyclic and collective with determination. The torque pedals became utterly unresponsive as the spin became uncontrollable. Vibration kept building as centrifugal force intensified.

Cabrillo and Robbins frantically ran through emergency procedures, trying desperately to bring what was left of the helicopter down easy. Michelle, nearly hysterical but holding on, called out the emergency checklist: "Harnesses; locked, Seats; up and Aft, Windows; jettison, Table; up and locked. Assume hard landing positions!"

Ali yelled over the roaring, whining port (left) engine, "Brace for impact! We're going in!" Michelle moved quickly, cinching up her harness tight. The rest of the crew and the Seals aboard did the same.

The cabin and cockpit section spiraled and rolled wildly into the desert floor, slamming into the ground with a crushing thud on its left side, auguring into the ground slightly to the left. Rotor blades whipping around, splintering into fragments in all directions! The pilot's door and outer bulkhead buckled inward into Ali's left side, crushing her left leg cruelly and pinning her in the seat.

The Second Rocket deals the final blow.

Coming around from the impact, Ali felt little at first. The crash caused severe injuries to her back. Lucky for her, those nerves stopped delivering pain to her brain, or else the agony from her leg would have been unbearable.

What was left of the tail section crashed into the ground away from the body of the helicopter, into the primary group of Yousef's ISIS fighters, killing them outright. The seal team in the back of the cabin was battered and beaten up by the crash but amazingly unhurt, well, mostly anyway.

Two of them were not so lucky. They'd been tossed violently out the cabin door along with Flanigan, the AWO, when the door and parts of the cabin structure were blown off and torn apart in the explosion from the second RPG.

Survivors recovered their senses from the shock of the crash, realizing they were still taking incoming fire from ISIS fighters. Seal training and muscle memory kicked in. They scrambled out of the cabin where the door had been on the port (right) side, clamored down off the wreck, and took up defensive positions around the twisted and mangled carcass.

Ruined bits of the once formidable machine lay scattered across the scene. Two of the guys from the Seal team began a rescue effort for Michelle and Ali. The others, who were now on the ground, some severely injured themselves, were fighting for their own lives.

Seals and Marines rescue wounded from the wreck.

Michelle's Rescue

Against initial orders to maintain radio silence, Specter 2 broke the mission protocols with a frantic request:

"Flight, Specter 2: Specter 1 is down, repeat Specter 1 is down, Specter flight and Garrison compound under heavy ground fire, request immediate air support, repeat, request immediate air support!"

"Specter 2, Fight: Understood request forwarded to air support coordination. Will advise ETA. Ready, Lighting is en route. ETA in ten. Status update to follow."

That day, the standby, "Ready Lightning," just happened to be Major Nash and his F-35… in case of emergencies like this one.

Specters 2 and 3 circled above the action on the ground, laying down withering cover fire from door-mounted M240 machine guns, covering the guys on the ground from the now-destroyed Seahawk. They circled in low enough for the Seals, Marines, and civilians onboard the other helicopters to rappel down to the ground.

The door gunners, still onboard Specter 2 and 3, continued to pour on heavy fire. One by one, they dropped into the malaise and engaged ISIS fighters, attacking the stricken helicopter and the garrison itself.

Ali was in pretty bad shape. The damage to her spine left her with no feeling in the lower half of her body.

She'd suffered severe lacerations to her thigh and was bleeding heavily from the wounds. The leg injury was so severe that it had turned 45 degrees inward from normal, with the bone protruding through the calf muscle and skin.

Flannigan, two of the other Seal team members, and Calabrese lay on the field from the explosion and fell to the ground when the door was blown off after the second rocket. All three men were killed outright from the rocket explosion.

Master Chief Brooks had fallen out as well, but wasn't killed outright. He'd been injured severely. He'd hit the ground, rolled over on his side, and started shooting, covering the guys trying to rescue the women.

The Ground Battle

ISIS pressed in hard, blasting with AKs. Firing more rockets at the wrecked helicopter and the other two still circling above, shooting full metal jacketed hot lead death at them. Luckily, these guys weren't very good shots. They missed the helicopter wreck! The surviving Seals, Marines, and contractors laid down a barrage of intense fire on the ISIS positions.

Their biggest fear was that they weren't carrying enough ammunition to keep up a prolonged fight. While this was happening, two guys from the Seal team were trying to pull Michelle and Ali out of the wreck! Deadeye was badly beaten up.

Michelle had broken her left arm and jacked her neck around, thrashing and flailing about in the cabin during the crash. She was shaken up pretty hard and nearly hysterical but otherwise coherent enough to respond to the guys pulling her out of the destroyed helicopter.

Muldoon, one of the seal team, a big burly dude, tried talking to her over the noise, nearly yelling at her, "Lieutenant, we're going to get you out; just take it easy!" He and Garcia, the Navy corpsman assigned to their unit, worked on getting her out and dodging incoming fire.

The two of them kicked out what was left of the bird's windscreen and climbed into the wreck, unstrapped her from the copilot's seat, trying to calm her down and avoid the incoming fire.

Garcia quickly looked her over with the intense ISIS fire hitting all around, trying to keep himself cool with her screaming. The two of them carefully pulled her out. Garcia yelled at Muldoon over the noise of the firefight, "Take it easy with her. We don't know how bad she is yet!"

He shot her up with a morphine syrette to ease the pain in her arm, calm her down, and keep her quiet while they worked ever gently to get her out amid the action going on around them. It's a hard thing to do with bullets and rocket-propelled grenades flying at you!

Their biggest fear was the wreck might catch fire or take another rocket before they could free both women. The battle raged on as these two guys worked to free them both. After they'd pulled 'Deadeye' out and laid her out next to the wreck, they used what little cover from it available as a shield from incoming fire.

Once they'd moved her to a safer place, 'more or less,' Muldoon put an m4 in her hands and told her: "Lieutenant, use this if you have to. Don't let them capture you! You understand what I mean, Ma'am?"

"Yes, I understand," she responded, shaking and frightened. She knew what he meant for an American woman and an officer. Getting captured would be worse than getting shot to death by ISIS terrorists.

The two guys then went to work getting Ali out, trying to avoid getting shot or blown up themselves. They both tried communicating with her. She was in decidedly worse shape than Michelle.

Muldoon yelled to her over the chaos: "Ma'am, are you alright? Can you speak?"

Ali responded with a painful but clear tone: "I can't feel my legs... Mm, ma, my back feels weird too!" Her response was groggy and half-conscious. She still tried to sound like the ever-strong woman in charge, though, and did her best to maintain a calm, forceful demeanor; "Hurry the fuck up, get me out of this thing!" came her painful response.

"Stand by; we're going to get you out as soon as we can!" The two looked at each other with grim, worried expressions. They knew from her response that she was in terrible shape, just by the look of that leg. Her muttering, slow responses, and what she said about her back, that it had probably been broken.

Her left leg had been partially crushed and pinned in by the weight of the wreck and the pilot's door. Ali was wedged into her seat between the center control console and the door frame. They both knew they wouldn't get her out before she bled out.

Garcia, the corpsman, rigged a tourniquet around her upper thigh and began trying to make her understand what she needed to do until they could get her completely out. "Ma'am, make sure the tourniquet stays tightly around your thigh to control the bleeding in your leg! Do you understand?"

"Ya, yeah, I understand," was her groggy response.

While they continued to work on getting her out, the battle raged on. The Seals, Marines, and POGs began running very low on ammunition. Muldoon gave her the same advice he had given Michelle: "Ma'am, do you have a sidearm with you?"

"Ya-yes."

"Take it out and be ready to use it if you have to. Do you understand what I'm telling you, Ma'am?"

"Yeah, I got it. Get me the hell out of this thing!"

"We will as soon as we can. Don't forget the tourniquet!"

"Okay, I got it."

The sentries in the compound had seen them go down from the guard towers near the garrison's main gate as the chaotic malaise exploded around them.

A group of POGs, some soldiers, and civilians at the compound loaded up Humvees and an MRAP with as many ammo cans as they could grab and mounted up to join in the fight. Others assisted with Michelle and Ali's rescue, and the other wounded, already started by Muldoon and Garcia.

ISIS was tenacious and pressed the assault forward. They sent the car bomb forward, attempting to break through to the gate next. It didn't get too far. The sentries in the guard towers fired two 50 caliber BMGs at the truck loaded with explosives. It blew up in a spectacular fireball before getting to the gate.

ISIS pushed forward anyway. They knew this would likely be their last day on earth. Yousef realized the end was near now. The helicopter crash diversion and their truck bomb tactics had failed, but he encouraged his men to push harder for as much death and destruction as possible to the infidels.

They pressed home the attack with all their strength, will, and faith that Allah would assist them. He didn't. They'd planned well and had the weapons supplied by their ISIS brothers, acquired from their command group through the Shura Council, and a network of fighters in Egypt and Libya. It just wasn't enough!

Russian-built RPGs and a cache of American-made stinger missiles, taken from the supposedly secret CIA station in Benghazi, Libya, during the 2013 attack on it by their brothers, at the American spy compound there, allowed Yousef, Nassir, and the others to shoot down Specter 1.

Gaining these weapons had briefly given them the edge, the ability to plan this attack and destroy the base, or so they thought. In the end, they were outgunned utterly.

The rancorous combat raged furiously. A spec in the early morning sky got closer and closer and began circling above. Spooky, The Angel of Death, the nickname for AC-130 gunships, had been flying standard patrol patterns in the area when the call was made for air support.

It was ordered to cover the troops fighting at Al-Tanf after the emergency air support call went out, and well - they got it! The 'Angel of Death' came calling, mercilessly strafing and blasting ISIS positions with heavy Gatling guns and 105 mm cannon fire.

Yousef, Nassir, and their cadre of fighters realized what was happening and that the end was near. All did what they could to prepare.

A chorus of "Allahu Akbar" went up as, one by one, they were picked off, blown apart, or torn to small bloody bits of flesh and bone by the Gunship's concentrated fire, the seal team's actions, and the troops from the base on the ground.

Michelle and Ali saw this as it went down! Men, ISIS men, torn to pieces by the terrible rain of fire from the Angel of Death, and their guys lying dead on the ground in the surrounding desert!

A sight neither of them should ever have had to witness, and neither would ever forget. Ali was less affected by it, though. She was too busy working that tourniquet, trying to keep herself from bleeding out and staying conscious. It affected her more later on.

Michelle was utterly terrified, even through the fog of the Morphine, and wouldn't recover from the experience. The sight of it stayed with her for the rest of her life, haunting her dreams and refusing to let her forget.

Multiple rounds of rifle fire hit Yousef. As he lay there in pain, bleeding out, he realized most of his brothers, his brave fighters, including his friend Nassir from Ramadi, were all already dead or near death. Some had slipped away, seeing the end coming, hoping to survive and fight on another day.

Laying there as life ebbed, he now understood what the Djinn in the desert had been trying to tell him! They would not only lose this fight but would lose many good Muslim fighters in a pointless, unwinnable, and unnecessary battle. A better strategy was needed. Now, it was too late for that.

The evil laughter of the desert Djinn was all he could hear now as consciousness faded and life ebbed away. His breathing became shallower and shallower. With his last few breaths, he recited the Shahada to himself as if to deny the Djinn satisfaction at his failure and death one last time and re-confirm his faith:

"لا إله ال الله محمدرسول الله"

"There is no deity but Allah, and Mohammad is his messenger."

Ali's Rescue

The battle came to its bloody conclusion. Some of the ISIS fighters had fled, but most lay dead around the battle scene. When they realized the fight was hopeless, a few of them skulked off into the hills and across the border back into Iraq.

Muldoon, Garcia, and two others began working feverishly to pull Ali out of what was left of the helicopter. They'd kept eyes on her as much as possible during the fight to make sure the tourniquet worked, and she didn't bleed out.

Garcia was the only medically skilled corpsman on the scene. He did a cursory assessment of the extensive wounds to Ali's twisted and battered leg and how best to deal with it. The tricky part was getting her spine and neck immobilized and positioned to avoid further possible spine or neck injuries.

Getting her out of the mangled wreck proved more complicated, especially since they were pretty sure her back was broken. The guys worked out how to stabilize what was left of the helicopter so it didn't shift, move, or collapse while they worked to get her out. Fortunately, the garrison had a good supply of field rescue equipment. A hydraulic rescue jack made quick work of it.

"Hold the jack steady, Garcia. One slip and this thing's coming down on all of us." Garcia pushed, grunted, and groaned, "I've got it. Just clear that floor panel." He muscled away the broken and shredded aluminum bulkheads, blocking access to Ali's seat. Luckily, the jack was available for emergencies like this.

Then, a portable cut-off saw cleared the remaining sheet-metal away from her lower body and left leg. The next problem was the floor and door post area that had crushed inward on her leg. Garcia glanced over at her. She looked pale and battered, strapped into the crumpled pilot's seat like she was. Garcia asked, "Ali, are you still with us?"

She responded, weak and groggy, "Yeah... still here. I wish I weren't, though." Garcia tried to sound encouraging. "Hey, no quitting on us now, Ma'am. We've seen worse, right?" "Speak for yourself, Garcia. I'm the one pinned in this wreck!" "We'll get you out, Ma'am. Muldoon, hand me the saw."

Muldoon passed Garcia the cut-off saw. Sparks flew as he cut through the twisted metal. Garcia yelled at him to check the tourniquet. Muldoon yells back at Garcia, "Jesus, this thing's a mess. What about her leg?" "Her leg? I'm more worried about her back. Look at how the seat is torqued over and twisted!"

Ali says through gritted teeth. "I can hear you guys, you know." "Good! That means you're still with us. Don't worry, Ma'am, we've got you." She adds, "It'd be great if you'd quit telling me that. Just get me the fuck out of it!"

Garcia slid a short spine board in behind her. Metal groaned as they leveraged apart the wreckage with a steel pry bar. "It's moving! Come on, almost there. Steady.. steady… okay, dude, go! Garcia inserts a full-length spine board behind her, bracing her back, working slowly and methodically.

Garcia asked her, "Lieutenant Commander, Ma'am, any pain?" Ali struggled to speak. "I don't know… still can't feel my legs. That counts as bad, doesn't it?" Garcia… "It's not great, but we're stabilizing everything. Just stay with us." Ali tried a faint smile. "Not like I have much of a choice." The last piece of wreckage shifted free with a loud, groaning metallic creak.

Muldoon and another marine, Davies, who'd been helping, gently maneuvered her onto the full spine board. "Easy now… Easy." Ali keeps talking to them. "You keep saying that. Just get me out of this thing! It's like I've been here forever!"

"You're almost out, ma'am. Hang tight." Garcia worked quicker now, splinting her mangled leg, wrapping it tightly. Then said, "Ma'am, I'm giving you another shot of morphine. This part's gonna hurt." Her eyes widen, and she freaks out a little… "Wait, what?" She winced, trying to bite back a scream as they straightened her leg. Muldoon winced alongside her and said, "Sorry, Ma'am. We had to do it. You'll thank us later."

Breathless from pain from the guys setting that leg, she quipped, "If I don't bleed out first!" Determined, Garcia tells her, "Not on my watch. Morphine's kicking in. Are you feeling it yet?" Groggier from the morphine, she comes back, "Feels… fuzzy. Still hate you, though, you prick!" Garcia grins. "I'll take that, ma'am."

Once Ali is safely out of the wreck, they carry her and Michelle towards a waiting Humvee, load them in the back, and climb into the seats in front of them. "Alright, guys, let's move! Get to the compound." In a soft but sarcastic voice, she tells them, "Thanks, guys. Even you, Muldoon." He grins, "That's the morphine talking, Ma'am. But you're welcome."

The Humvee roared to life and headed for the garrison's medical dispensary. As they pulled away, the wreck still smoldered. Overhead, a lone F-35 buzzes the scene. Major Sam Nash was in the ready lightning, flying "high guard" for his favorite Navy pilot, Ali.

Chapter 5: Charlie Mike!

Medevac: Dust-Off

The decision had been made to fly them both back to the ship on Specter 2. The pilot got on the radio and talked to the ship's surgeon. "Specter 2, flight; how do you read?"

"Flight, Specter 2; you're five by five."

"Specter 2, flight; Loading wounded for medevac now. ETA approximately two hours."

The ship's surgeon was listening to the radio calls and wanted to talk to Garcia directly to get a better idea of Ali's, Michelle's, and the conditions of the wounded, and how best to deal with moving them to the ship; "Specter 2, Flight; patch me through to the Corpsman."

"Rodger that." Specter 2's pilot radioed on to the base dispensary,

Garcia was there, treating Cabrillo, Robbins, and the other wounded inside the compound's medical spaces. Their radio operator patched him through, "Garcia, the Ship's surgeon, wants to talk to you."

"Patch him through. Go ahead," Garcia responded, "Garcia here."

"Garcia, Commander Slatter. Status report on the two pilots' conditions."

"Sir, the copilot was shaken up severely, thrown around in the crash. Her left arm is broken in several places; no compound fractures, though. Nothing sticks out of her anywhere. We've splinted her arm, but she's also displaying possible concussion symptoms."

"She was pretty hysterical; She kept asking about the pilot's condition. She seemed worried about her, mostly. I gave her morphine to calm her down. She's quiet now and more or less stable."

"And the pilot? What's her condition?"

"She's in pretty bad shape, sir. Her left leg is broken in several places and crushed below the knee—multiple compound fractures. I've done what I can for it, but she'll need x-rays and surgery for sure as soon as possible—apparent compression fractures in her back. We've immobilized it as best we could."

"She told us before we shot her up with morphine that she couldn't feel her legs and wasn't able to move them. She said her back felt weird, too. I assumed the worst. We immobilized her spine and neck, pulled her out of the wreck, and moved her, Lt. Robbins, and the others to the compound's dispensary. We also have a couple of the Seal team and several other wounded marines, sir."

"Belay Specter 2 transport. We'll send an Osprey Medevac. ETA 40 minutes."

"Aye, sir, we'll get them ready. Is there anything else you want us to do?"

"Send the less critical cases on Specter 2 and return them to the ship soonest. Proceed with standard protocols for all wounded. Stabilize them as much as possible."

"Understood, Garcia out."

Ali's condition was severe, but stable for the time being. Garcia spent most of his time with her, monitoring her condition and waiting for the 'Dust-Off' medevac to arrive. After unloading the gear and supplies from Specter 2 and 3, the less severely wounded were carried into Specter 2. Casualties were zipped into body bags and loaded into Specter 3.

The Dust Off flight arrived in short order. Ali, Michelle, and the other critically wounded Seals and Marines were carried onboard on stretchers. Garcia accompanied them to brief Petty Officer Sanchez on everyone's conditions.

The medical crews on these aircraft typically comprise two personnel: a flight surgeon, in this case, Warrant Officer Hayes, a P.A., and Hospital Corpsman HM2 Sheila Sanchez. Hayes added his two cents to the load-out, "Garcia, assist Sanchez with these patients. Ensure we get a thorough status pass-down on them all."

These two were regular medical crew on medevac missions like this one from the ship. Besides being trained as emergency medical professionals, these crew members were also trained to handle emergency water landings and water survival skills like non-medical flight crews train for.

Garcia answers him back with, "Yes, sir. I've detailed everything I had the ability and equipment to do for them."

Hayes kept asking him questions, "How are these people doing? What's their condition now?"

"All of them are stable for now. Vitals are within normal range. Keep a close eye on the pilot, though. She's critical but stable. She's lost a lot of blood. We didn't have any of her blood type available. I've been giving her morphine to keep her from trying to move around and to keep her quiet."

"Copy that. How long since the last dose? Sanchez, see if we have AB-negative blood onboard."

Garcia replies, "Since we brought her to the compound a couple of hours ago."

"Okay, we'll keep her sedated until we get back. Sanchez, administer another round of meds to keep her comfortable as needed during the flight."

"Yes, sir. I'll monitor all of them as needed. We might have some turbulence. It might make things a little challenging to keep everything connected."

"Copy that. Double-check and secure as needed."

"I'll take care of that, too, sir."

"Also, advise the ship's medical team of her status and specific requirements on arrival and make it as seamless as possible."

"Yes, sir, I'll call in status before arrival."

The Osprey took off from the LZ (landing zone) on a northwest trajectory, heading towards the Syrian coast. Its twin tilt rotors roared as it lifted off at a nearly vertical angle, churning up swirling dust clouds behind. The aircraft flew low over the desert terrain, avoiding radar locks from SAM missiles that Syrian forces used efficiently.

It was a smooth flight back, but it took them over some hot territory. The flight crew kept eagle eyes out for missiles, paying close attention to the missile warning system. The countermeasures system was on hot standby, just in case.

Dust-off's pilot poured on the speed, avoiding potentially dangerous areas as best he could, and returned to the ship in record time. Hayes wanted more patient updates an hour into the flight: "Corpsman, status report, please."

"Yes, sir, Lieutenant-Commander Cabrillo is still quiet. All her vitals are normal."

"What about Robbins?"

"Her vitals are stable, sir. Stats are also normal. She's been coming around a bit. She keeps asking about the Lieutenant-Commander."

"Keep an eye on them both."

"Understood. Will do, Sir."

The ship was offshore, about 50 miles off the coast of Larnaca, Cypress, steaming towards the Syrian coast. It had been running its regular patrol pattern, doing flight operations, drills, and other training. Dust-off neared its security zone around the ship, so the co-pilot started his radio calls for approach instructions:

"ATC, Dust Off 1, Approaching security zone, request landing instructions."

"Dustoff-1, ATC, report current position."

"ATC, Dust Off-1, current position; 34.50.00° North, 33.00.00° West, altitude 8,000."

"Copy Dust Off-1. Welcome home. Standby for approach instructions."

"Copy ATC. Standing by."

Back On the ship

The dust-off flight was about 630 direct air miles to the ship. An Osprey isn't as fast as a jet, but they're much quicker than cargo helicopters like CH-46s or CH-53s.

The conditions of their patients, especially Ali, called for as quick a trip as possible, so the flight crew "poured the coal to it" and made it back in record time. It was still a two-hour flight—a long one by Osprey standards.

Sanchez had been checking everyone's vitals and general condition throughout the trip. She made her final checks before they arrived. The Aircrew announced over the comms that they'd arrived.

"Prepare for landing, entering the ship's traffic pattern. Med crew, initiate standard landing procedures."

When she heard the comms crackle through her headset, Sanchez readied everything, securing the medical bay as needed, and strapped into her seat.

Hayes had already strapped into the jump seat behind the pilot and copilot. The radio call went to the ship's traffic control for landing instructions. Since this was a Medevac flight, the call sign was 'Dust-Off-1'.

"Gitmo traffic, Dust Off-1, entering approach for landing. Request priority clearance. Standby medical teams for patient transfer."

The ship's ATC responded, "Dust off-1, ATC; cleared for spot 5, standard approach, follow landing signals, med team notified and standing by."

"Copy ATC, standard approach - spot 5, follow landing signals."

"Copy Dust Off-1, welcome home. We have you in sight. Proceed as directed."

"Copy ATC, spot 5, standard approach. Thank you."

Flight deck crews readied for recovery. The deck had been cleared and checked for FOD (loose flotsam and jetsam), so nothing got blasted around by the Tilt-Rotor's twin engines or getting sucked into one and destroying it. Dust Off-1 entered the pattern, wheels down, in landing configuration. It was an easy touchdown on Spot 5. The LSE waved them down on the deck, signaling to shut down the engines. The blue-shirt crew set the wheel chocks and applied tie-down chains.

Medical Spaces

Dust-off's shutdown was complete. The front access door opened, and the crew exited. The aft cargo door opened, and the rear ramp dropped. The dust-off's med crew, assisted by the ship's medical department (flight deck white jersey crew, red crosses on the front and back), began hustling the wounded off the aircraft.

They carried Ali, Michelle, and the rest to the starboard side catwalks and down the access ladders to the deck below. The white shirt crew directed the walking wounded to follow behind, heading for the medical department for treatment.

Carrying stretchers through a ship is dicey: starboard side forward, port side aft, through the passageways, all the while dodging bulkhead opening knee-knockers (shin-level edges.) Not a simple task!

Back in medical overflow, patients were moved into beds, except for Ali. The ship's medical staff took her directly to an exam room.

A Slow Recovery

Days and weeks following Ali's surgery seemed to pass slowly to her. It took a few days for the effects of the pain drugs and anesthetic to fade away. Hospital staff made sure the pain meds were enough to keep her comfortable, though. They gradually took her off of the opiate drugs. It didn't take long before the stiffness and soreness set in.

Conway came by as usual a couple of weeks into her recovery, making daily rounds and checking in on her and his other patients. Greeting her in his typically friendly but clinical manner.

"Good morning, Lieutenant-Commander. How are you feeling today?"

"Good morning, Sir. I'm getting the feeling back in my legs, Captain, but it's a mixed blessing."

Her response puzzled him, so he asked, "What do you mean?"

"Well, on the one hand, I'm relieved my spine is healing, and I'm getting the feeling back in my legs, but now I'm feeling a little pain and stiffness. I'm still pretty sore."

"How would you describe the pain you're feeling? On a scale of 1 to 10, how bad is it?"

"It's a dull throbbing soreness most of the time. It's not too bad until I try to move around or go to the head; then, it's really painful. I'd give it a 2 or 3 most of the time, and maybe an 8 when it really kicks!"

"Okay, I understand. I'll have the nurses increase your pain meds for now."

"Thank you, Captain, but I'm tired of feeling like a drugged-out zombie, sir."

"Alright, we'll try something milder for the inflammation and soreness. Is there anything else?"

"No sir, that's fine. I'm doing well otherwise. It's just tiresome laying in this bed all the time."

"You should be able to get up and move around a bit in a couple more weeks and start physical therapy shortly after that. It'll be six to eight weeks before you're healed up enough to start any serious physical therapy. As for anything else, if you're serious about getting back into good enough physical shape to return to flight status, that will be more like several months."

She grimaced, realizing it would be an uphill fight to return to flying again. She'd overheard several conversations about a potential transfer to Bethesda's Walter Reed National Military Medical Center. This was news she didn't want or need to hear.

She knew it probably meant the end of any hope of returning to her squadron, maybe never being able to fly again.

"Why would I not want to do that, Captain? I fully intend to return to flying. That's what I do!"

Conway knew she would be anxious about it, but he didn't want to give her any false or unrealistic expectations.

"Well, Lieutenant-Commander, as you know, aviation physiology flight standards for Aviators and NFOs are strict. I'm not sure that your leg, back, or, for that matter, your ankle and foot can handle the stress of physical training like that again."

"I'm not a flight surgeon, though. If that's your intention, all I can say is to be prepared for failure. Moreover, if you damage your lower leg and ankle again, you may lose it! It's being held together with pins, plates, screws, and bone grafts. It will never be as good as before the crash, ever again."

"I understand what you're telling me, sir, but I fully intend to try."

"Well, in that case, all I can offer you is the best of luck. But be aware of your limitations and don't push it too hard. Full recovery will take six to eight weeks before you can begin physical therapy. There are a lot of other jobs you can do for the Navy, you know."

"I'm aware of that, sir, but I have invested far too much in my flying career to just give it up."

Then the Captain dropped the next surprise on her. "By the way, I also need to let you know that you'll be transferring to Bethesda as soon as you're able to get around better for extensive physical therapy. We aren't well equipped for that here. If anyone can get you back to flying condition, they can."

She lay there, stubbornly thinking to herself, "I'm goddamn not giving up on flying! I can't let it end like this! That damned crash! I can't let it end this way!"

She was still angry with herself, feeling it was her fault that it went down like it did.

Long about then, a certain Dr. Carver, a visiting psychotherapist, dropped by to see her, "Lieutenant-Commander Cabrillo?"

"Yes, that's me; what can I do for you?"

"I'm Dr. Carver. I'm a psychologist for the Naval Hospital. 'He didn't specify which one.' I'm here on a temporary assignment. I was wondering if I could chat with you if you feel up to it. Would that be all right?"

Now Ali was thinking: "Oh shit, here it comes, shrink time!"

"Yes, I suppose that would be all right; come in. What did you want to chat about?"

"I just wanted to see how you are feeling and if you're up for a chat about your experiences and recovery since the crash. Would you mind talking with me a bit about it?"

"Yes, I suppose that would be all right. I'm a little tired, but we can talk for a while."

"Okay, great. I'll start by asking how you're sleeping. Are you having any trouble trying to sleep?"

"Well, no, not really. That's pretty much all I've been doing since I was pulled out of that helicopter wreck in Syria! I've only recently been taken off the heavy drugs they've been feeding me since then."

"Okay, then. Since you have just recently become fully aware again, have you noticed any changes in your moods or emotions since the crash?"

"No, not really. Other than I'm pretty sick of lying in hospital beds and really want to get up and start moving around again."

"Well, that's normal. Dr. Conway has told me you still have a few weeks before you can do that, though. It's just going to take a little more time before that happens. Do you ever feel on edge, or do you seem to be easily startled?"

"Dr., I have only just begun feeling somewhat normal again; I'm on edge about not being able to return to flying, though."

"What did Conway say about that?"

"The Captain said I may or may not pass the aviation physical and flight standards again. My injuries may not heal adequately enough to allow that much strenuous physical training for me to return to peak condition."

"I see. What are your feelings about that?"

"I told him I was going to try no matter what! Flying is, after all, my profession. I absolutely will try to get my flight status back."

"Did he say anything else?"

"Only that I should be prepared for a negative outcome if it comes to pass and offered me his good wishes."

"Have you experienced any feelings of guilt or shame related to the accident or injury?"

"Well, I'd be less than honest if I didn't admit to feeling guilty about the others that didn't make it. It was my command, and they didn't come home, so yeah, I feel responsible for their loss!".

"It wasn't your fault, Lieutenant-Commander. There was no way you could have expected, planned for, or avoided an attack like that."

"Hey! I survived, even though I'm a mess right now! They didn't, and most had families with young children, so I feel really guilty about that!"

"It wasn't your fault you were attacked and shot down in a combat action. Try to put it into a fair perspective for yourself. Okay, just a couple of more questions."

"Okay, fine. What else, then?"

"Do you feel detached or disconnected from others?"

"Are you kidding! I'm laid up here in this hospital bed. I couldn't get out of it myself for several weeks now, trying to recover and come back to reality and life! Does it seem to you I have attachments?" Ali was decidedly irritated by the question. The aggravation raised her blood pressure and made her leg throb. That just pissed her off more.

"Okay, fine. Try not to get upset; they're just standard questions to help me understand your feelings. There's nothing to get angry over."

"Fine, what else do you want to know?"

"Just one last question. Have you had any thoughts of self-harm or suicide?"

"Not yet, but there are a few people I feel like shooting at the moment!" She looked directly at him with an irritated expression and daggers in her eyes to drive home the point! If an angry look and a scowl could kill, Carver would have been lying on a slab in the hospital's morgue.

Ali's annoyed expression told Carver he might have tapped a nerve by asking the wrong question at the wrong time.

Startled at her response, he backed off and ended the conversation. "Well then, I think that's all I need for today. I'm sure we'll be chatting again soon. I'll let you get some rest now. Thank you for your time, Lieutenant-Commander."

"You're welcome. Goodbye, Doctor." Irritated on the inside, she was thinking something more like, "Don't come back, asshole! What stupid ass questions to ask somebody laid up in a hospital bed like this!"

His inane questions made her aggravated and offended at his obvious psycho-babble and seemingly silly questions. That hot Latin temper got the best of her.

"Thank god I'm starting PT soon - I so want to get out of here!"

Ali's Condition

The ship's chief medical officer, Commander J. D. Slater, was a neurologist by specialty, not a surgeon, although he often did simple surgeries when needed. He wanted as thorough an assessment of Ali as possible.

Everything they could do for her back injuries had been done. She was already immobilized and sedated to keep her from moving around too much. Slatter decided he wanted to speak to her directly and ask questions, so he ordered her brought around.

"Get her prepped for a full exam, get her blood work going, check her vitals, and start a plasma drip. What were her stats when you picked her up? I want to do a comprehensive evaluation of her condition. Let's get a full set of X-rays on her back and that leg as soon as we're done. Bring her around enough so we can talk to her, but make sure her pain meds are working."

"Yes, sir, understood; I'll prep her and bring her around. The medic from the Seal Team reported she'd lost a lot of blood. We didn't have her blood type available on the Medevac aircraft or at their base. She needs a transfusion ASAP. AB negative."

The doctor was visibly disturbed by this news. "Damn! I don't think we have any in our stock either. Check and see if we have it, anyway. If we don't, contact the bridge, have them pass a request over the 1MC (ship's public address annunciator) for any ship's crew, squadron personnel, or embarked Marines onboard with AB negative to lay to medical for blood donations."

"Belay that. Just let me know if we have any AB-negative in our stocks. If we don't, I'll talk to the captain myself."

"Yes, sir, I'll get it done right away!"

Since Sanchez was on the Medevac flight, she was already familiar with Ali's stats and condition. She injected her with a shot of naloxone to bring her around enough for the doctor to talk to her and the other medical staff.

HM2 Burdick checked their blood stocks for type AB-negative blood. Sanchez noticed Ali coming around and tried talking to her to bring her out of it faster. "Ma'am, Lieutenant-Commander Cabrillo, can you hear me, Ma'am?"

Ali was still pretty groggy from the drugs. She was having trouble trying to speak. She moaned, trying to respond: "Mm, Wh-who are y-you, w-where am I, w-what's happen-ing-g!".

Sanchez tried letting her know what was going on. "Ma'am, I'm HM2 Sanchez; you're back on the ship. Your helicopter was hit by rocket fire and crashed in Syria."

"You were hurt pretty bad. Try to be calm. We're going to take good care of you now. Commander Slatter, the ship's doctor, wants to talk with you. He'll be here shortly. Are you having any pain anywhere?"

"N-no, no pain. I-I can't move; why can't I move anything? Cr-crew, how's my crew?" "It's okay, Ma'am. We have you immobilized until we find out how serious your back injuries are."

"Lieutenant Robbins is here, too. She's hurt, but she'll be okay. I don't know about any of the others from your aircraft. Just take it easy and rest for now. You're in good hands. Commander Slatter, the ship's doctor, will be here shortly."

Sanchez was relieved that Ali was not feeling any pain. She thought sure Ali would scream and freak out if she were experiencing any. Fortunately, the pain meds were working as hoped. Ali was groggy and barely conscious but was responsive.

Corpsman Burdick returned after checking blood stocks and quietly told Sanchez, "We don't have any AB-negative blood in our stocks." "Okay, thanks. I'll let Commander Slatter know."

Slatter returned, fast walking into the room. "How is she doing, Sanchez? Status, please?"

"She's awake, just barely, sir. She understands when I talk to her. She is still pretty out of it, but responsive.

It's a negative on the blood type, sir. We don't have any AB-negative on hand."

Slatter; "Okay, thank you, Sanchez. I'll handle that. Now, let's see how you're doing, Ms. Cabrillo." He peered over the exam table at her and spoke slowly. "Ms. Cabrillo. Lieutenant-Commander Cabrillo, can you hear me? How are you doing?"

She spoke slowly, sleepily, slurring her speech in a sort of hung-over voice: "I-I'm still alive, I g-guess, n-not m-my b-best.

I-I can't feel my l-legs, w-why I can't feel my legs?" She was getting worried now about that and started getting a little worked up over it.

"Don't worry about that right now. Your back was injured severely in the crash. You'll regain the feeling in your legs soon. You'll have to give it time to heal before that happens. In the meantime, we have you immobilized to keep from damaging anything else till things improve."

"H-How long?" She asked."

Slatter answered in a vague, non-answer manner doctors often use when they don't want to tell patients everything or aren't sure:

"That's hard to say at this point. You've lost a lot of blood. We're going to give you a transfusion as soon as we wrangle up a few pints of your blood type. You've got a rare one."

"We'll check with the other ships in the task force and see if they have any available. If not, we'll get the crew to donate for you. In the meantime, take it easy and try to rest. Sanchez here will take good care of you. I'll be back to check on you soon."

Sanchez; "Yes, Ma'am, I surely will. Let me know if you feel any discomfort or pain, and I'll adjust your meds."

"O-Okay, I-I'll try to."

Slatter was off to his office and called the ship's bridge directly to speak with the CO, Captain Manning. The bridge watch picked up and passed on the message, "Captain sir, Ship's Doctor for you."

"Captain Manning."

"Captain Sir, this is Commander Slatter."

"What is it, Slatter?"

"Sir, as you probably know, we brought two pilots back on the Medevac flight today. One of them is Lieutenant-Commander Cabrillo. She needs a transfusion of a blood type we don't have as soon as possible."

"Unfortunately, we have none of it in the Ship's medical inventory. I want to check the other ships for it. If no one has any readily available, we'll need to look for donors in the ship's company, the embarked marines, and the squadron personnel."

"Very well, Slatter. We'll contact the other ships in our group. Senior Chief O'Toole or one of his people in the CR division will let you know. We'll pass the word to the crew and the squadron personnel anyway to donate."

"Yes, sir. Thank you, Captain."

The doctor hung up and headed back to the medical overflow. He hurried along, looking at other patients from the Medevac flight. Since Michelle was the next most seriously injured, he started with her, directing her to be moved, "Burdick, move Lieutenant Robbins to exam room 2. I want to look at her next. What's her status?"

"Yes, sir, she's stable; her vitals are normal. She keeps asking about Lieutenant-Commander Cabrillo's condition. She seems very concerned about her."

"Never mind that; it's fine. They're friends, and she's the Lieutenant- Commander's Co-Pilot. Just let her know the Lieutenant-Commander is all right for now. What's her blood type?" Slatter had been hoped she might be AB-negative as well, but 'No Joy,'

"She's O negative, Sir."

"Shit! Okay, never mind, move her to exam room 2 and prep her for a complete workup. I'm going to check up on the Lieutenant-

Commander again. After I finish with her, I'll be right there to examine Robbins."

"Yes, Sir, right away, Commander."

The Doctor dog-trotted back through the clean passageways and white-painted bulkhead openings over the blue and well-polished linoleum on the decks. Sanchez was still tending to Ali in exam room 1.

"How's she doing, Sanchez?"

"She's stable, but her blood counts are very low. She needs that transfusion as soon as possible, sir." They kept their conversation low-key, just above a whisper, so Ali couldn't hear clearly what they were saying. She wasn't really listening.

Ali was only half conscious and still more or less in a drug-induced lethargic state, hungover from the morphine and other meds they'd pumped into her since being pulled out of the wreck. About then, another petty officer walked in.

"Commander Slatter, sir, there's a Major Nash here to see you. He says it's about Lieutenant-Commander Cabrillo."

"Tell him to wait; I'm busy right now!" Slatter mutters, a bit irritated he'd been interrupted.

"Sir, he says it's about the blood Lieutenant-Commander Cabrillo needs. He's got the same blood type."

"Okay, great, get his ass in here right now!"

"Yes, sir, right away."

Burdick was a little startled at the Doctor's response. Slatter Had never been pushy or angry, except for now. He showed Nash into the exam room:

Slatter; "Nash, I understand you have type AB negative blood. It's pretty rare to find someone with it. Are you sure about it?"

Sam replies, "Well, Commander, that's what I'm told, and that's what it says on my dog tags, so I have to assume that's right. Anyway,

Lieutenant-Commander Cabrillo and I are old friends. I want to help if I can."

"Okay, good. Sanchez, get him prepped for a blood draw right away." "Yessir, Major Nash, please sit, and I'll get you hooked up."

"Anything for Lieutenant-Commander Cabrillo!"

Ali was still groggy and out of it, but she'd heard the conversation and recognized Nash's voice. After his last comment, she grumbled quietly in a slightly sarcastic tone. She spoke in an irritated, grumpy voice people use when they're really sick, lethargic from medication as she was, or like somebody who's been awakened abruptly:

"S-Sam, I'll r-remember you s-said that."

"Ali! Are you okay? I heard about the crash. We've all been worried about you. Are you okay?"

"Sh-shut the f-fuck up, Sam. Let me rest. You always talked too much, Marine!" They all laugh hysterically at her grumbling sarcasm.

Slatter chimed in with his learned medical opinion: "Well, that's a good sign! Still got some sarcasm in there, eh, Lieutenant Commander?"

"Yes-s Doctor-r, I guess s-so."

"Okay, that's enough talk for now. Take it easy and rest. Petty Officer Sanchez will stay with you and give you some of Nash's blood. You'll feel better after that. Just take it easy and rest. I'll be checking in on you later."

"Major Nash, we can only take a pint of blood from you at a time; otherwise, you'll end up lying there next to her yourself. Pass the word on to your squadron mates and the others. We still need blood donations from personnel with AB-negative blood. Tell anyone with that blood type to lay to medical on the double."

"Yes, sir, I'll pass the word around as soon as I'm done here."

The doors to the exam room swung open. HM-3 Farnum entered, looking for Slatter with news from the other ships.

"Commander Slatter, sir, Master Chief O'Toole called from CR division to tell us two other ships in our task force have AB-negative blood on hand. Some have reported donors are also available if needed, as you requested. They'll have their on-hand stock flown over to us as soon as they get it ready."

Slatter was a little more at ease at this news; "That's great! Did they mention how much they could provide?"

"No sir, they didn't say how much they had."

"Okay, we'll just have to keep looking for more if it isn't enough. Sanchez, I'm going to look at Lieutenant Robbins now. Make sure the Lieutenant-Commander gets the additional blood units as soon as they arrive. We'll take a good hard look at that leg after she's had the transfusions and recovered a little."

"Yes, sir. I'll handle it."

Robbins' Condition

Slatter quick-timed it to exam room 2, swung the door open, walked in, and looked straight at Petty Officer Burdick with a demanding scowl; "Burdick, did you get those x-rays done I asked for yet? Is she awake? What are her stats currently?"

Intimidated by Slatter's demeanor, Burdick thought it was really out of character for him; "Yes, sir, I got everything you asked for. She's awake and can understand you, but keeps complaining about having severe headache symptoms. She also keeps insisting on finding out Lieutenant-Commander Cabrillo's condition."

"Okay, good. Let's have a look at her."

Burdick could tell Slatter was more than a little stressed by the current demands on him: the intense activity with Medevac patients, the search for blood for Ali, and demands for diagnosis and treatment of these patients, especially Ali.

Her injuries were the most severe. Slatter wanted to ensure he overlooked nothing. He pressured himself to be as thorough as possible. Now he focused his attention on Michelle.

Slatter looked over her x-rays and addressed her. "Lieutenant Robbins, how're you doing? Are you feeling any better? How's the arm? Are you having any pain anywhere?"

"My head hurts. My arm isn't bothering me that much. Can I have something for my headache? It hurts. I can't even rest or sleep. Please, something, anything!"

"Sure, Lieutenant, in a bit. But first, let's check out a few things, okay?" Michelle couldn't argue with him, the Ship's doctor and a Commander to boot. She answered in frustrated and aggravated tones,

"Okay—I mean, I guess so, Commander."

"Did you lose consciousness?"

"Only after Sanchez shot me up with morphine or whatever it was!"

"Trust me. You wanted that. Okay, do you remember what happened immediately before the crash?"

"We got hit with a couple of rockets. The starboard engine caught fire, and we went into a flat spin after we lost the tail rotor. After that, it's a blur."

Slatter questioned her further. "Have you been feeling dizzy or lightheaded?"

"I don't know about that. All I know is I've got a splitting headache that won't go away!"

"Have you been experiencing nausea or felt like you might have to vomit?"

"No, not so far, Sir."

"Are you sensitive to light or noise?"

"I don't think so, but I've been out of it for a while. It's not loud here, so I can't say about loud noises. The lights aren't bothering me."

Slatter continued, "Are you having trouble concentrating or remembering things?"

"Only after we started spinning in during the crash. It was a blur after that."

"Okay, Lieutenant Robbins. Petty Officer Burdick will give you something to help you rest and relieve your headaches. Your x-rays look good, and your arm should heal fine. I'll be around to check on you later. By the way, the Lieutenant-Commander is resting comfortably and has just received a blood transfusion from Major Nash."

"Okay, thank you, Doctor." Michelle scowled when she heard Nash was the one who donated blood for Ali's transfusion. She didn't enjoy hearing that name. She knew Nash and that he had a thing for Ali. It made her jealous at the thought of him even talking to her, let alone giving her some of his blood!

Slatter noted her symptoms, commenting on her medical record: *'Possible concussive T.B.I. symptoms noted. Continued observation ordered.'*

"Burdick, I'll be with Lieutenant-Commander Cabrillo if anything comes up."

"Yes, sir."

Slatter was out of Exam Room 2, quick-timing it back to Exam Room 1. He walked in as Sanchez was already giving Ali Major Nash's donated blood. "Status Perez: Have you finished the transfusion?"

"Yes, sir, I'm almost done. We also received more blood units from the other ships. I'm adding those next."

"Outstanding, Petty Officer Perez. What's her current condition?"

"She's sleeping now, sir. Her stats are normal, and her blood count is improving."

"Okay, great. Let's look at those X-rays again."

Slatter scrutinized the X-rays like an eagle scanning a panoramic landscape of some empty savannah, looking for prey. He studied anything and everything on them for indications: severity of fractures, bone fragments or displacement, and damage to nearby blood vessels or nerves. "Let me know when she wakes up. I want to get a closer look at those leg injuries. Let her rest for now."

"Yes, sir, I'll watch her closely."

Slatter was more than a little concerned Ali might develop a severe infection around those leg injuries, since the fractures around her ankle and foot were so severe. He sat at a desk in the exam room, trying to decide if he should perform this complicated surgery or not.

He wasn't so confident about doing the surgery needed to set her broken left leg and attempting to repair that shattered ankle and foot on the ship himself or just having her Medevac'd off the ship to the hospital at Sigonella or even on to Landstuhl Germany.

Slatter's biggest fear was that if he did the surgery himself, there was a better than even chance his patient could develop a severe infection, maybe even die from it. The thought weighed heavily on his mind and made him even more cautious about doing the complicated surgery himself.

He felt he needed to inspect Ali's wounds again and scrutinize those X-rays more closely than the first time he'd seen them. "Petty Officer Sanchez, let's have another look at those X-rays. Fetch them for me, please."

"Yes, Sir, I'll get them."

Perez handed him the X-rays. He stuck them in the viewer, flipped the switch to light it up, and studied them closely. Sanchez thought he was taking quite a lot of time looking at them, but passed it off as him being thorough. She wasn't about to question him over it, anyway. She figured he was working out his approach for the surgery if he even tackled it himself.

After scrutinizing them again, working through the what-ifs, Slatter had seen enough and decided he needed to ship Ali off to Sigonella. They had an actual orthopedic surgeon and, even better—a much better equipped surgical facility.

"Perez, pass the word to Warrant Officer Hayes to join us for a short meeting to discuss the Lieutenant-Commander's condition."

"Right away, sir," was her response.

She located Hayes, the Ship's flight surgeon. In reality, only a PA, but a very good one. They returned to the exam room to discuss Slatter's concerns about Ali's condition.

"Bryan, I asked for you to join Perez and me to discuss Lieutenant Commander Cabrillo's leg and my decision regarding her surgery."

"Well, sir, I can more or less assume you will send her to Sigonella. Neither you nor I have ever done anything like this before. Setting those bones and pinning her ankle and foot together will be complicated. I'm not sure we could successfully do this complicated surgery here without the risk of her having greater complications later."

"Moreover, as I'm sure you know, there's a considerable risk of infection developing that we might not be able to deal with here on the ship. She needs an orthopedic surgeon for this, and considering her other injuries, CAT scans and MRI imaging wouldn't be a bad idea either."

"Indeed, those were my conclusions. Well, that settles it, then. Let's prep her to be medevac'd to Sigonella first thing tomorrow morning. Since you both are the medical crew on the medevac bird, I'll leave those details to you."

After Slatter completed his tests and made his conclusions about her condition, they rolled her back to the medical overflow ward and parked the gurney next to Lieutenant Robbins. Michelle was awake and watched as they transferred her to the bed beside hers. "How is she?" she asked. "I'll live, I guess, for now," was Ali's groggy but typically terse response.

"Ali! You're awake! Thank God you're all right!" Michelle was ecstatic that Ali seemed okay and was near her.

"I didn't say I was all right, Shel. I feel like shit. I'm tired, and everything hurts except the stuff I can't feel right now."

"What do you mean? You're here, and you are alive! I'm so relieved; I was afraid you were dead!"

"That still might happen, Shel, so don't get too excited yet; I'm really fucked up and exhausted right now. Can we talk later if I ever feel better again? I need to rest now."

"I'm sorry, Ali. I didn't mean to bother you."

"It's okay, Shelly. Thanks for worrying about me. Now I'm going to sleep for a while."

"Okay, Ali, rest. We'll talk later. Maybe tomorrow."

"Yeah, maybe."

Michelle was overjoyed to see Ali still alive, even though she was clearly in pretty bad shape. She was still troubled about her. Michelle was just like that where Ali was concerned.

Ali's Departure

The ship's medical staff took her early the following day. The medical ward was actually a troop space used as an overflow ward. The Corpsmen lifted her from the rack while she slept quietly on a gurney, preparing her for the trip to Sigonella. They didn't want to disturb the others, so they tried to keep it quiet.

Slatter and his medical staff had told no one of her transfer off the ship for surgery. He didn't want to wait, knowing it would affect her odds of keeping or losing her leg. The move to the exam room stirred her from sleep.

"Wha, what's going on? What are you guys doing now?"

"Don't worry, Ma'am. Commander Slatter wants to talk with you. We're just prepping you."

"For what?"

"The Commander will explain. Don't worry; it's nothing serious. Just relax."

That freaked her out. Whenever she heard those words, it always meant bad news. A short while later, Slatter entered the exam room with a quick, forceful gate, almost like he was marching. He was not a strict officer, but he always wanted his staff to feel like he was the strong leader type.

As if he even needed to do that. He was a doctor and a Navy Commander. Nobody was ever going to challenge him or argue with him. His medical department staff thought he got off on the power trip as the ship's medical officer a bit too much. They understood his B.S. Some grumbled about it. Most just ignored it.

"Lieutenant-Commander Cabrillo, how are you feeling this morning?" "Well, I'd probably be better if your staff hadn't woken me up so early; I'm okay, I guess; why? What's so important at this hour," was her strained, drowsy, drug-induced, and slightly irritated response.

"I wanted to speak to you about your condition and be sure you understand what's happening. We want to make sure you get the best care possible. You need critical surgery to repair and treat all the bone damage in your lower leg, ankle, and foot before they heal in ways that may cause you problems later. There's also a significant danger of infection."

"We don't have the facilities or the surgical talent for such a surgery here on this ship. Just understand that we need to get you to Sigonella as soon as possible so they can do it for you. We don't want to risk you being disabled or losing your leg." Now Ali started getting worried. "W-well, okay, wh-what's the plan, then?"

"I've already cleared it with the CO and Commander Deering. You're relieved of all squadron duties. You'll be medevac'd off the ship shortly. Commander Deering will see that your gear and records are forwarded to Sigonella." "W-why? Won't I be coming back?"

"You'll be there for a while, and there's a possibility they'll ship you back to Bethesda. We can't be certain when or even if you'll be coming back. Perez here, and Warrant Officer Hayes will accompany you on the flight. You'll be departing within the hour."

"I see. That's Wh-why the early reveille."

"Petty Officer Perez will get you prepped. She and some of the other staff will take you up to the flight deck for departure. Good luck to you, Lieutenant-Commander."

Perez added, "Don't worry, Ma'am, I'll take care of you. Just relax and try to rest some more."

Ali carped a little, her apprehension building, "Yeah, like I'm going to do that now."

Perez and the others moved her onto a stretcher, entered the elevator next to the medical department, and headed for the flight deck. It creaked and groaned a little as it rose. Dust-Off 1 was already prepped and ready to fly as soon as they loaded up, and the rest of the flight crew boarded.

Perez and Hayes made sure everything else was ready to go. Long about this time, the ship started its usual daily routine: Reveille at 06:00, breakfast for the officers and crew, 'muster on station,' a daily ritual head count of everyone on board the ship at 07:00. Then, at 08:00, the 1MC public address system crackled out a shrill whistle from the bosun's pipe. Then, the announcement, "flight quarters, flight quarters," carried throughout the ship.

The flight deck came alive with activity. Dust-off's crew went through their start-up procedures. The engines on the Tilt-rotor came to life, whining and spinning up with a high-pitched whistle to a deafening roar as the power built up. Then, Lieutenant-Commander Ali Cabrillo was off on her second medevac flight. This time, headed for surgery at Sigonella Air Station's base hospital.

Slatter made the intelligent decision to send her there. He'd radioed ahead and given them the whole situation—her condition, the severity of her wounds, and his diagnosis. Not being an actual orthopedic surgeon himself, he didn't want to risk it. "Dust-off," the same Tilt-Rotor Osprey that brought her and the others back from Syria, took off and headed for Sigonella, a flight of a little over 360 miles from the ship's position.

This type of aircraft takes about three hours to fly to Sicily from where the ship was located. The medical crew, Warrant Officer Hayes, the Ship's PA, and HM2 Sanchez were the same two who'd taken care of her during the Syrian medevac flight. Both were regulars on dust-off missions and intimately familiar with her condition.

It was an easy flight to Sigonella, a routine trip, except for the cargo. Lieutenant-Commander Ali Cabrillo. She couldn't help but feel like this was it for her Navy flying career. She felt like she was abandoning her friends, her squadron, and their CO. She lay there quietly, sobbing a little to herself.

"God dammit, why did this happen? I can't believe this is how it'sgoing to end!" She couldn't shake the feeling that she wouldn't return and would never wear gold Aviator wings again!

Sanchez sensed her uneasiness and said, "You all right, Ma'am?"

"I'm okay, just kind of depressed and frustrated at all this bullshit."

"I understand, Ma'am. We'll do our best to get you back in the saddle soon. Just rest for now, *Soy tu chica de casa. ¡Te cuidaré bien!*"

Ali appreciated Perez's personal touch in Spanish. Although she didn't speak it much now, it made her feel a little less isolated to know there was someone from home looking after her, with a personal touch like this. She responded to Perez in as personal a way as she could. *"¡Gracias prima, lo aprecio mucho!"*

Back in the ship's medical ward, the other patients from the crash and the battle in Syria, including Michelle, stirred and awoke to the noise of the 1MC public address system with that rude, abrupt startle you get when awakened by something or someone interrupting a sound sleep.

The first call was, "Reveille, Reveille—All hands heave out and trice up—breakfast for the crew!" Later came "Muster on Station" and then "Flight Quarters, Flight Quarters." Startled awake by all this noise, Michelle rolled over to check on Ali, wanting to see if she was awake yet and ask how she was feeling.

Shocked that Ali wasn't beside her any longer, Michelle half-thought about getting up and finding out where she was. She moved to get up. That turned out to be a mistake. Her head pounded as soon as she tried to sit up.

The headaches she'd been suffering since the crash never really went away completely. When the meds wore off, the headaches returned with a vengeance. Today was no different. She stayed in the bunk and decided it might be better to ask the corpsman about Ali when he came by to check on everyone.

A little while later, Burdick came by, doing his morning rounds. Michelle had to ask: "Corpsman, what happened to Lieutenant-Commander Cabrillo? Where is she?"

"We Medevac'd her off the ship early this morning, Ma'am."

"What! But why? What happened? Is she all right?"

"She needed emergency surgery. The ship's doctor opted to send her to Sigonella. They have more extensive surgical facilities and an orthopedic surgeon on staff, so the Commander sent her there."

"When will she be back?"

"I don't know, Lieutenant. It depends on how well the surgery goes. I suppose. She'll be laid up there for a while. There's a good chance she might never be back."

"What, what do you even mean? That can't be! I, we need her here!"

"Like I said, it depends on many things."

Michelle reeled at the thought that Ali might never return to her and their squadron. It upset her and made her tear up, thinking she might never see Ali again. "Oh no, that just can't happen!"

Burdick tried to encourage her a little. "If we can't get you over your headaches and back to your regular duties soon, there's a good chance you might join her there anyway, Lieutenant."

"What! Why would you say that?"

"You'll have to discuss that with Commander Slatter, Ma'am. I can't say for certain." Burdick wondered, "Why is she so worried about her?"

Chapter 6: Sigonella Naval Hospital

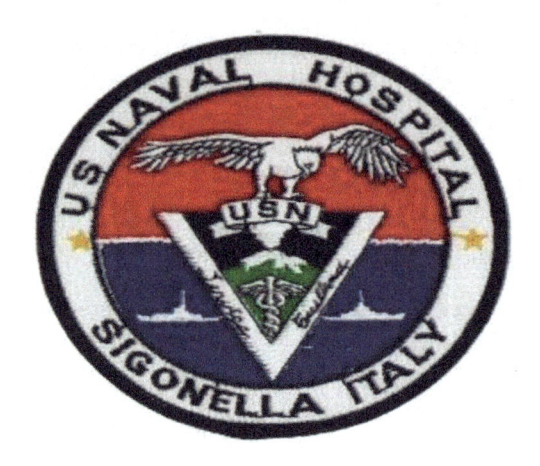

The First Surgery

The Sigonella facilities were far more extensive than the ship's medical department was equipped for. More importantly, they had MRI imaging, extensive diagnostic capabilities, and a dedicated surgical team with an orthopedic surgeon on staff. Ships like the Gitmo are well-equipped to handle combat casualties, but delicate surgeries like Ali's, not so much.

Amputations are more the normal range of their skills—a more typical approach to extensive fractures like hers. The ship's doctor, Commander Slatter, felt strongly that Ali needed MRI imaging before any surgery could be considered or done successfully. He didn't want to tackle that work himself, anyway.

Captain Conway, the chief surgeon at Sigonella, was an actual orthopedic surgeon. He began evaluating Ali's injuries right after she arrived, ordering a full suite of MRI imaging to assess the true extent of her injuries and any associated damage like bone fragments where the bones had been shattered, crushed, or displaced and any undiagnosed tissue damage nearby, blood vessels, nerves, and muscle.

He and his surgical team initially assumed they could handle everything based on Slatter's diagnosis and their follow-up tests after she arrived. It had been initially thought she would be transferred to Landstuhl, Germany, or directly to Bethesda, stateside in Maryland.

The Captain wanted to give a reasonable effort before making a final decision on the transfer. He was more concerned with her leg injuries than the spinal injuries. They'd already immobilized her spine after the crash and on the ship, had her in traction, and kept her sedated enough to keep the pain under control.

His diagnosis after imaging and extensive tests was pretty much what he'd already discussed with Commander Slatter, the ship's doctor. She had 3 compound fractures. One above the knee and two below the knee, and ankle and foot had been mangled and crushed by the impact.

He also felt Ali's condition was severe, but manageable for his surgical department. He held off on another transfer to Landstuhl for now. The Captain wanted to tackle this patient's case himself. Others on his staff were not so confident in his abilities, but he was an O-6, so that was that.

Conway determined the approach he and his surgical team would follow. In this case, multiple compound fractures, with open wounds where bones had broken through the skin. His biggest concern was infection they mightn't be able to control. The plan was for immediate surgery to stabilize the fractures and align the bones properly before they were mended out of proper alignment.

A Corpsman and one of the staff nurses settled her in after the flight when the Captain dropped in to see her, to let her know what to expect before and after the surgery.

"How's she doing? Are you guys getting her settled in? Is she awake?"

"Yes, sir, she's awake, but a little out of it."

"That's fine, as long as she can hear and understand what I'm telling her."

"I cc-can hear you j-j-just fine, Captain."

"Good. Ms. Cabrillo, I came by to chat with you and let you know theseverity of your condition and what we've planned for your surgery."

"Okay, w-what is it then?"

"Well, I wanted to tell you what to expect and make sure you understand your situation clearly."

"Okay, sir. What do you want to tell me then?"

"Your ship's doctor, Commander Slatter, may or may not have fully briefed you, so I'll give you a full rundown now: You've sustained multiple compression fractures in your spine. The ship's medical staff has already dealt with that, immobilized you, and done the appropriate corrective treatment."

"It will take some time, but you should expect to regain sensation in your legs and lower extremities over the next few days to weeks. However, the time that will take varies on a case-by-case basis." "Considering your youth and excellent physical condition, your prognosis is optimistic."

"Your left leg, well. That's a more serious matter. You've sustained multiple compound fractures above and below the knee. The impact of the crash has crushed your ankle and foot."

"We think we can save them, but you aren't likely to have full mobility again. Now, I must stress there's a genuine possibility you might still lose your foot and portions of your lower leg."

"If the damage is too severe, or complications arise, you may lose them. We'll do our very best to save them, though, before discussing that any further. Do you understand fully what I'm telling you, Lieutenant-Commander?"

Feeling Despondent now, she replied, "Yes, sir, I understand completely. You're telling me I'm likely to be crippled either way! Is that right, Captain?"

"Surprised by her response, he tries to sound positive, "Well, I wouldn't put it so starkly, but essentially that's correct. You'll be able to walk and get around fine. Either way, you're not likely to run again if you're concerned about that."

He felt like he had to be frank and honest with her now:

"You're not likely to fly again for the Navy either, but you'll be functional otherwise. Life will be more or less normal for you. In reality, you're pretty lucky. This kind of crash doesn't paint a pretty picture for survival and recovery for anyone."

"Sorry, Captain, but that doesn't make me feel better about it."

"Try not to worry too much about it right now. Let's see how it goes. Your surgery is scheduled for 0800 tomorrow morning. I'll see you then. Till then, get some rest."

"Thanks for your honesty, Captain. I appreciate you stopping by personally to explain the situation to me."

Feeling more fearful now that her career may be over, she didn't hold back. She lay there, feeling sorry for herself, sobbing quietly and agonizing about her uncertain future.

"What the hell will I ever do now?"

Early the following morning, hospital staff wheeled her to the pre-op room. There, they were met by Dr. Claudio De Luca, the anesthesiologist. He greeted her warmly in a heavy Italian accent: "Signorina Cabrillo, how are you feeling this morning?"

Drowsy and half-awake, Ali replies, "I'm pretty tired."

De Luca gave her a little grin, teasing her a little, and whispered in her ear, hands expressing in a typical Italian manner as if sharing a secret: "Don't-a worry, I'm-a gonna give-a you the good stuff! It will help-a you go right back to sleep, presto!"

A dark-haired, tall, good-looking Italian. His English skills weren't the best, but De Luca was flamboyant and jovial. He loved to flirt with the nurses and his female patients. De Luca's IV anesthetic drip didn't take long to kick in. Half asleep already, Ali just drifted off into that deep, unconscious anesthetic world these drugs send us to.

A short time later, Conway came by pre-op to check on her. "Ms. Cabrillo, how are you feeling?" No answer. The attending nurse tells him, "She's already under, sir."

Conway replies, "Okay, wheel her into the OR, and let's get started."

"Yes, sir."

Dutifully following the order, the prep nurse rolled Ali down the hall into the elevator, one of those large, made-for-a-gurney medical elevators hospitals and medical facilities have, and pressed the button for the lower-level ORs. It creaked and groaned on its way down, like an old factory freight elevator, with that hard, abrupt feeling of hitting the brakes when it stopped.

The doors opened automatically, out of the elevator and down the hall towards the OR. The surgical team was already set up and waiting. Conway, De Luca, the anesthesiologist, and the rest of the surgical team. The OR was sterile, well-lit, and well-equipped, with state-of-the-art surgical instruments and monitors.

The tension and anticipation were heavy as Conway's team embarked on this complex and intricate surgery. He began with a quick review and pep talk,

"Okay, before we get started, let's briefly review the plan for the Lieutenant-Commander's surgery: goal number one: Repair her shattered lower left leg, ankle, and foot—the limb of a professional and highly valued Naval Officer and Aviator. Considering the extensive damage, we'll do everything we can to repair it as well as possible. Goal number two, of course, is getting her back on her feet and back to a normal life as soon as possible, hopefully even getting her back to duty status. Let's get it done."

The OR buzzed with controlled chaos, each playing a crucial role in this delicate and complicated surgery. The OR nurses draped Ali's leg with sterile sheets, exposing only the affected areas.

Conway gazed critically at her legs in a moment of chauvinistic male admiration, thinking to himself,

"Man, this woman has such beautiful legs! What a shame this happened."

While that went on, the scrub nurse meticulously checked and arranged sterile, gleaming instruments under harsh surgical lights.

De Luca started the anesthesia flow, maintaining the proper depth to keep Ali unconscious and unaware during the intricate reconstruction procedure. IV lines ran to infusion bags containing de Luca's anesthetic cocktail and plasma. "She's-a ready for induction. Let me know when you're-a ready to start. Do you have-a any specific preferences, Captain?"

No, we're fine. Let's get to it." Ali's steady breathing told them that behind the array of equipment and surgical talent lay a human life—a valuable officer who put her trust in Conway's and his team's skills. Monitors displayed vitals. A steady rhythm reassured Conway, De Luca, and the others that she was ready.

His gloved and steady hands hovered over the exposed and damaged leg like a sculptor contemplating a block of marble. He worked with deliberate and skillful incisions, opening the areas around the fractured bones and damaged tissue, moving through layers of skin, muscle, and blood vessels.

De Luca, the attending nurses, and the OR techs kept close eyes on the monitors, assisting as Conway called for instruments. Feverishly, they worked the process, passing instruments between themselves and him, watching the monitors, each playing their part in the intricate and stressful surgery.

After he'd opened her up, Conway carefully and thoroughly examined the x-rays and MRI scans of her leg, ankle, and foot again, studying closely the extent of the fractures and bone damage. With a concise plan in mind, He went to work, making incisions along the side of her leg, exposing fractured bones, and performing debridement (cleaning out decayed and dead tissue).

"Starting debridement. Saline and suction, please?"

"Ready, doctor; saline is on your left, suction in place."

"Perfect. Let's make sure we get all the devitalized tissue. Keep an eye out for bleeding. Let me know if it gets excessive."

"Will do. Hemostatic agents are ready if needed."

"Great, keep those handy. Adjusting incision angle for deeper tissue."

"Understood. Suction is ready. Monitoring fluid levels."

Then came a stabilization process. Conway realigned her broken bones into their proper positions, aligning the intact portions of her shin bone, pinning them together with surgical screws and plates and other bits of internal surgical devices.

"Ready to start stabilization. Assessing alignment of tibia and fibula fragments. Bone reduction forceps, please."

"Forceps, Doctor. Imaging on screen for your reference."

"Using reduction forceps to align fragments. Apply traction to maintain alignment."

"Traction in place. Holding alignment steady."

"Good. Placing internal fixator. Drill and guide pins, please."

"Drill and guide pins ready."

"Drilling holes for guide pins now. Maintain alignment!"

"Understood. Keeping the field clean, monitoring for bleeding."

"Guide pins in place. Plates and screws."

"Plates and screws ready."

"Placing plates over fractures and securing. Verify correct alignment with the X-ray."

"Alignment looks good. X-rays indicate proper position."

"Tightening screws and securing plates. Check for stability and ensure there's no significant movement."

"Checking stability now. Fixation appears secure."

For the splintered and crushed bones, Conway delicately removed the fragments and cleaned and stabilized the surrounding tissue. Since so much of the bone in her leg, ankle, and foot had been splintered or crushed, donated portions of bone were grafted into place for what was not possible to save—hours passed.

The fractures and damaged tissue were meticulously aligned and stabilized, pinned, and screwed together. Shattered bones were pieced back together wherever possible, like a complex jigsaw puzzle, meticulously reconstructed, and torn ligaments repaired.

Once everything was aligned, pinned, and bone grafts stabilized, Conway applied morphogenic proteins in a gel form directly to affected areas to stimulate growth factors and enhance natural bone-forming processes. As the surgery neared its end, Conway carefully sutured the layers of tissue and closed the incisions, sealing her leg with sterile dressings.

Conway exhaled and sighed heavily. Relieved the surgery had gone as well as it did.

"Excellent. Closing incisions."

"All dressings and sutures in place. Closure complete."

The last knot was tied, and a collective breath of relief filled the OR. Conway took a step back to survey the work. Satisfied with everyone's job, he ordered: "Wrap it up. We'll apply the cast later. Take her to recovery."

"Yes, sir!" came the collective response from his OR team.

Later, in recovery, the attending nursing staff monitored her closely as she came out from the anesthetic, pain meds, and antibiotics ready to be administered as needed. The recovery nurse began trying to pull her out of the anesthetic by talking with her: "Miss Cabrillo, Miss Cabrillo."

"Wha-what…"

"Wake up, you're all done!"

Coming out of that anesthetic drug haze, Ali thought to herself, "Oh shit! Fucked up! Mouth dry." It seemed to her like she was exhaling dust in every breath. Her first words were, "Wa-water, some water," just above a lethargic whisper.

"Not yet; just hold on until you come out of it more!" After a few hours in the recovery room, they rolled her back to her room, still keeping a pretty close eye on her. The floor nurses monitored her for signs of infection or complications. Pain management was now the order of the day.

Physical Therapy

The following morning, a local Italian doctor, Dr. Angelo Modena, the hospital's resident physical therapist, started his initial evaluation.

"Well, Signorina Cabrillo, let's get started. Today, we focus on your range of motion and flexibility in your knees and hips. These exercises should help you regain feeling in your legs faster. We start with knee flexes, hip rotations, flexors, and abductions." He put her through her paces, working slowly but building up as she went.

"How does that feel? Any new or more severe pain?"

"I'm still a little stiff and sore. I can't tell if it's new or old."

"Don't worry. If you can feel anything, that's a perfect sign. I understand you will be transferred to Bethesda Naval Hospital soon, so we will try to finish as much work as possible with our time. Okay?"

"Excellent, thank you for that! I'm so sick of laying in this hospital bed. I wanna scream sometimes!"

"Ah, that's normal; most people feel that way. After they remove the cast from your leg, if you are still here, we will work on your ankle and foot. We'll start with ankle rotations, toe curls, and range-of-motion exercises. They will help you with flexibility."

"Okay, now we give you a soak in our whirlpool jacuzzi bath, too. That should help to reduce pain and swelling and improve your circulation."

"How will you ever do that with that cast still on my leg?"

"We've got a fix for that. We wrap it in plastic to keep it dry."

"Okay, whatever you say." She was skeptical but figured, "What the hell? He must know what he's doing."

So, she just played along and enjoyed the jacuzzi sessions. Several days passed with more physical therapy and jacuzzi baths each day.

By this time, she could get out of bed and move around in a wheelchair, but she still couldn't yet walk on crutches.

Conway came by her room, making his usual rounds, but earlier this morning, to give her the word about her transfer: "Good morning, Ms. Cabrillo; how are you getting along today? Feeling any better?

"Yes, sir, I'm feeling closer to normal again. At least I can get out of bed and go to the head by myself now."

"Good. I'm glad to hear you're feeling better finally. I have some news for you today."

"Really, what's that, Captain?"

"You're being transferred to Bethesda for further therapy and rehabilitation. We've gone as far as we can go with you here, and you still need a few weeks to heal before that cast comes off."

"I've been getting excellent reports from Angelo on your progress, and we feel you are in good enough shape to transfer now."

"Oh, okay, so when will that happen?"

"First thing Monday morning, a medical transport will fly in, so you and a few other patients here will fly to Andrews Air Force Base on that flight. From there, they will take you to Bethesda."

Chapter 7: Bethesda

Walter Reed Medical Center

Ali's ride to Bethesda that following week arrived early Monday morning. An Air Force C-17-AE medevac transport fitted out as a flying hospital. The Sigonella ambulance crews moved and loaded her and a few other patients headed for Bethesda onto the aircraft.

Once everyone was brought aboard, the Air Force med-techs settled them in for the flight. While all that was happening, ground crews refueled the big hospital transport for a quick turnaround. When that was completed, it took off and headed for Andrews Air Force Base in Maryland.

The flight to Bethesda was easy and uneventful. The medical team onboard monitored everyone closely, tracking vitals and medication. Touchdown at Anderson was smooth and easy. The flight crew understood their mission well: safely deliver medical evacuees, some in life-threatening conditions.

Patients didn't need any unnecessary bouncing around to complicate things. The crew did their best to touch the ground as light as a feather. The big hospital bird rolled to a stop. The cargo ramp came down at the rear. Med-techs carried patients off the plane to waiting buses and ambulances, where they would be taken to whatever medical facility they were bound for.

More Therapy

Ali's transport was bound for Walter Reed, where most onboard were headed. After a quick trip from Andrews, the transport pulled into Walter Reed and unloaded Ali and the others who'd been critically injured in some godforsaken place in the deserts of the Middle East.

Once she'd settled in, therapy and rehab entered its second phase. Dr. Radcliff, a noted physical therapy specialist, dropped by her room to introduce herself and get Ali started with more advanced therapy.

"Lieutenant-Commander Cabrillo? May I come in?" The doctor asked.

"Um, sure, why not? Come in," Ali replied.

"I'm Dr. Portia Radcliff; I'll head up your physical therapy and rehab during your stay with us."

"By all means, come in!" Ali thinks to herself, "Finally, some action!"

Radcliff asked, " Are you up for getting started today? We've got a lot to do."

"Oh, thank you! I'm so ready for it!"

"Good, we'll get you down to the rehab clinic shortly and get started. Have you had any issues or additional pain since you left Sigonella?"

"No, just sleep and boredom! I want to get back on my feet and back into shape as soon as possible!"

"Well, one thing at a time. Let's get you back to standing and walking before you even think about running!"

"I suppose you're right. I'm just really frustrated by all the crap that's happened!"

The bone grafts and plates in her leg and foot were precarious at best. The pins holding her ankle together likely wouldn't hold under the heavy stress of what she had to do to get back into shape.

Radcliff suspected they couldn't handle the stress of prolonged physical training and the pounding of running and working out. She was optimistic when talking with Ali, but she knew her situation and the odds of success.

"We'll continue where you left off at Sigonella until the cast comes off your leg. That will still be a couple more weeks, so let's get going. In his reports and treatment plan, Dr. Modena indicated you were doing knee flexes, hip rotations, flexors, and abductions with him. How was that going?"

"I told him I was still sore and stiff."

"Okay, we'll start working on those exercises until your cast comes off."

Radcliff and Ali worked on Modena's original plan for several weeks until her cast was removed. Radcliff wasn't there when it happened. When she saw it, she was shocked to see so much purple bruising after a month and a half in a cast. She did her best not to seem overly concerned about it in front of Ali. Radcliff went on as planned, hoping it would clear up.

"We'll start your phase two therapy today with soft tissue and joint mobilization. We'll work your leg muscles using hands-on manual manipulation to get them working and flexible again. Please let me know if you experience any discomfort or pain."

Ali was a little apprehensive about the pain issue but tried to forget it and put up a determined face in front of the Doctors.

"Okay, let's get to it!"

"Great, let's get started then. This technique will help break up scar tissue, improve blood circulation, and enhance healing. Let's work that ankle with some flexibility and rotation exercises."

Radcliff worked her ankle through different ranges of motion. Ali's grimaced facial expression told Radcliff she was experiencing pain and trying to hide it.

"Are you sure you aren't experiencing any pain?"

"A little, but it's not bad."

"Well, tell me if you do! We don't want to damage anything any further. We'll adjust the treatments if it's a problem for you."

During these sessions, Radcliff closely tracked Ali's progress, monitoring her pain responses and adjusting treatments when and where needed. Gradually, she regained strength, and her stability improved.

Radcliff started having her do weight-bearing exercises to get her back up and on her feet, walking and bearing weight on her injured leg. Next, Radcliff had her do balance, coordination, and strength exercises.

Ali progressed quickly as therapy went on. Her work and effort paid off after a short time. She regained more function and mobility every day. Fractured bones were healing. The crushed and splintered bones in her ankle and foot were healing, but still a source of concern to Radcliff and the other doctors.

Therapy progressed and involved more active exercises to improve strength, flexibility, and balance—exercises to strengthen calf muscles and other injured areas. Activities to improve coordination, proprioception, and awareness of her body position and coordination were added to workouts.

Bad News—The worst happens.

During the final stages of physical therapy, Radcliff's plan for Ali focused on functional activities and returning her to regular daily routines. Radcliff had her do exercises to improve walking and jogging. Ali wanted to start slow running and get her endurance back up as quickly as possible. She got to where she could walk well but still had discomfort and pain trying to run.

Radcliff warned her repeatedly in no uncertain terms:

"Lieutenant-Commander, I urge you to slow down. Your lower leg and ankle are precarious at best. If you push it too hard, you're likely to have severe problems and a lot more pain. You're taking an enormous risk by pushing it so hard too soon!"

"Okay, I know, I know, Doctor! I'll try to take it easier and slow down."

"I'm serious now! You could damage the repairs done to your leg and maybe even lose it!"

"I'll keep that in mind, Doctor."

Ali wasn't listening very much to her. Radcliff's harping just aggravated her more. Her stubbornness and determination became obsessive. Overdoing as she was and trying to run on it when she really shouldn't have even attempted to was taking its toll, hitting her with a vengeance one afternoon during therapy and workouts.

She started fast jogging and slowly running, trying to ignore the pain in her ankle and lower leg. The pain became unbearable. She doubled over, let out a painful scream, and tumbled head over heels to the ground. The pins in her ankle gave way, and everything just came apart.

Ali let out an agonizing scream and tumbled head over heels. The repairs holding everything together came apart from the stress she put on them. Radcliff saw it as it went down, cursing under her breath.

"God-Dammit! I knew that would happen!"

Radcliff immediately ran to help her, punching the hospital's emergency number into her phone for help. Lucky for her, she was already at the hospital. When emergency doctors got to her, they found she'd broken the tibia bone again running on it.

Repairs to her ankle had failed to heal correctly. Worst of all, looking at the damage, they realized her body was rejecting the bone grafts. They took her directly into emergency surgery to repair the damage a second time and save her leg, but without the bone grafts, that wouldn't happen.

The underlying cause was a persistent infection nobody detected, which was worsening. At that point, the decision was made to amputate her lower leg to save her life. When Ali came out of the anesthesia, she wasn't yet aware of what had happened. Recovery room staff whispered between each other so she wouldn't hear what they were talking about. It made her suspicious; she started asking questions,

"He-Hey, what the hell's going on here?"

"Take it easy, Ma'am; the doctor will be in to see you shortly. You've had emergency surgery. You broke your leg again during your workout. The doctor will explain what the surgery was for."

She knew something horrible had happened and, fearing the worst, the anger built up over this 'no answer' response.

"Why are you all whispering about me? What the hell's going on here? God-dammit, answer me!"

She shrieked in fear and disbelief when she realized that her lower left leg was now missing. Recovery room nurses and attendants jumped in to get her under control, giving her a sedative to calm her down and keep her from thrashing around.

"Calm down, Ma'am, calm down."

"Bullshit! Somebody better tell me what the hell happened!"

The 'no answer' answer from the staff just pissed her off even more. Her shrieks were now screaming and cursing. Finally, the sedative kicked in. She quickly faded out. Ali was still pretty worked up, though, realizing her lower left calf and foot were gone, cut off, amputated without even an explanation from a doctor!

Her battered and now half a leg was wrapped and covered strangely in a way she hadn't seen before. Ali was devastated, knowing it would likely end all hope of flying again. Sometime later, she came back to consciousness, and the surgeon who had performed the procedure was there,

"Lieutenant-Commander, are you with us?" he asked.

"I suppose so; at least, most of me is." She responded tersely, not trying to hide her grief and anger over the whole thing.

"I'm Doctor Schnieder; I'm the surgeon who performed your procedure."

"You mean you're the feckless asshole that cut off my leg? Well, I can't say I'm pleased to meet you then!"

"I understand your anger, Ms. Cabrillo, but we couldn't save your leg. The pins and plates in your tibia and ankle gave way. You put excessive stress on them working out. I understand from talking with Dr. Radcliff that she warned you about overdoing it."

Schneider continued, "That's not the whole thing, though: during the surgery, we found that the bone grafts weren't healing properly. Your body was rejecting them, and you'd developed a severe infection. So, there wasn't any other choice. It likely would have come down to you losing it anyway, or it would have taken your life in the end."

"So, what now? When do you decide whether I'm out of the service?"

"That will be up to the medical review board in San Diego."

"What do you mean? I'm already at the National Military Medical Center. Why would you even send me there?"

"Balboa Hospital has the best amputee clinic in the Navy. They'll fit you with a prosthetic leg and foot and get you back on your feet again before the medical review board makes that determination."

"They'll review your physical condition, ability to perform your regular duties and psych evaluations."

"What the hell! Psych evaluations! Why would that even enter the picture? I don't need a damned shrink to tell me if I'm fit for duty or not!" Schneider's comments pissed her off a lot. She knew what that meant, and it just aggravated her even more.

"It's not really about that, Ms. Cabrillo. You've been through a very traumatic experience. You'll be evaluated for traumatic stress and whether you're able to cope with the further hardships of duty in the future."

"That's not what it sounds like to me!"

"I'm just trying to tell you what to expect, particularly if you're thinking about trying to stay on active duty."

Ali took a step back mentally and paused before she said: "Yes, I understand that, doctor. I apologize for yelling and screaming. Sorry if I seem angry, but I'm fucking angry! I'm aggravated, upset, frustrated, and fucking angry! I mean, look at me! I'm grounded, and probably for good!"

"I understand your feelings, but you must realize, and I'm sure you've heard this before, that you're fortunate to have survived a crash like that and then recover from your injuries as well as you have, even if it's not 100 percent."

"You're right; I've heard it all before. It doesn't make me feel better about any of it, especially now!"

"Your transfer orders will come through in several weeks. In the meantime, try to get some rest and let that leg heal properly. I'll drop by occasionally to check in on you while you are here with us."

"Great, thanks for your concern, doctor."

The doctor didn't care for her sarcasm, but He understood she had just lost a limb after already dealing with combat trauma, a fractured spine, surgery on her leg, and working for weeks trying to get it and herself back on track, only to lose it in the end.

Chapter 8: Balboa and Beyond

As the doctor said, things would happen quickly; Ali's transfer orders took several weeks to process before she was notified that she would ship out and head for San Diego's Balboa Naval Hospital Amputee clinic. The flight there was like the trip from Sigonella to Bethesda, an Air Force medical transport.

This time, she was awake and sitting in a passenger seat, not lying on a stretcher on the bare aluminum deck of the transport. It arrived at Miramar, where Ali and the other patients on board were offloaded at the passenger terminal and transported to the Naval Hospital.

After everything that had happened, the crash, recovering from her back injuries, months of rehabilitation and physical therapy, only to end up ultimately losing her lower leg and foot, it came down to this: Lieutenant-Commander Althea 'Ali' Cabrillo was headed for the renowned Balboa Amputee Center.

A state-of-the-art facility known for its cutting-edge technology and experienced staff. It was the perfect place to be fitted with a state-of-the-art prosthetic limb, undergo more physical therapy, and still have a shot at getting back in the cockpit.

After arriving at the center, Dr. Emil Rodgers, a highly skilled prosthetist, greeted her.

"Good afternoon, Lieutenant-Commander; I'm Dr. Rodgers. I'll be overseeing the fitting process for your new prosthesis."

Rodgers took his time explaining the entire procedure to her, making sure she understood every step.

"We'll start by evaluating your condition, taking some measurements, and assessing the condition of your leg's skin and muscles. That information will be crucial. We need it to make your prosthesis a perfect fit and give you a more normal working and living routine."

"Fine, but how does it perform during physical workouts? I still want to get back to flying, you know."

Rodgers wasn't going to fall into that trap so easily. Promising her something she wouldn't likely be able to qualify for.

"One thing at a time, Lieutenant-Commander, one thing at a time."

"All right then, let's get to it!"

"First, we'll scan your leg to create a digital model. That'll guarantee a precise and accurate fit. We'll use it to fabricate a socket that interfaces between your leg and the prosthetic lower limb and foot."

"How long will all that take?"

"The scans won't take long, but the actual fabrication of the interface could take from a few days to a few weeks, depending on how fast the company that makes them can turn it around."

"Okay, let's get going then!"

"After the interface is built, we'll start you on physical therapy and get you walking and standing normally."

Several more people entered her room about then. Rodgers chimed in with introductions for his physical therapy team.

"Ms. Cabrillo, I'd like to introduce you to your physical therapist, Dr. Marcel Bélanger. He's from Montreal, Quebec, in Canada. He's one of the best therapists for amputees in the world."

"It's nice to meet you, Lieutenant-Commander. My team and I will work with you to get you up to speed as soon as possible." Ali spoke fluent French, so she offered him a greeting she thought he might enjoy: "*Au plaisir de vous rencontrer, Dr.*"

"Thank you, but let's be polite to everyone and stick to English while we're here if you don't mind."

"Okay, sure, that's fine. Sorry. What's first?"

"Well, for now, since we're waiting for your interface appliance, we can start strengthening your thighs, hips, and core. That should help your balance a lot. When the appliance is ready, Dr. Rodgers will do several fitting sessions with you to ensure it fits correctly and works as advertised."

When it arrived, Ali slid it on the stump of her left calf, and the fitting process began. Rodgers made meticulous adjustments to align it perfectly with her body's mechanics. This took a good bit of time and patience. They worked together to fine-tune the fit.

Physical therapy sessions continued. Bélanger and his team focused on strengthening the muscles surrounding her residual limb, improving her gait and overall balance. They introduced various exercises and techniques to help her adapt to the new prosthetic.

Sam Visits Ali

Weeks of work went by. Ali's progress was remarkable. With each passing day, she grew more confident with the new prosthetic leg, walking on it, doing a slow jogging run, and even climbing stairs, but awkwardly. Physical therapy sessions were tough, but she needed them. She had to regain her independence and pursue her goal: putting her gold wings back on and flying again.

Ali got a little help from an old friend as she navigated this new and challenging path. A knock came on her hospital room door.

"The door isn't locked; come in."

"Hey Ali!"

"Sam! Sammy Nash! What are you doing here?"

"I've been trying to track you down since they moved you to Sigonella from the ship. After we got back from the Med, I got new orders. I'm now squadron CO for a new F-35 Squadron out of Miramar. When I heard you'd been moved here from Bethesda, I made some inquiries, and well, here I am, at your service, Ma'am," pleased he'd finally found her.

"Sam, how does a major make squadron CO?"

"I'm not a major anymore! The assignment called for a Lieutenant Colonel. And well, I got promoted!"

"That's great! Congratulations, Sam! But I wish you had told me you were in San Diego first. I hate for you to see me like this."

"C'mon Ali, like what? I know what happened. It doesn't matter to me. We're old friends, and well, I'm crazy about you, anyway. I've been worried about you since then! I just wanted to check in and see how you were doing and offer to help if you need anything."

"Thanks for that, Sam, but you didn't need to. I'll be okay."

She teared up a little at Sam's concern for her. After all, as far as she was concerned, they'd only been friends in the past. Sam realized she didn't believe what she'd said to him.

"It sounds like you don't think you're alright or will be, Ali. Why do you have doubts?"

"I'm afraid they'll ground me for good! I want to return to flying and do what I was trained to do! What I've always only wanted to do."

"Do you feel you can even pass the flight standards again?"

"Well, I'm sure going to give it a hell of a try! I need to get past this prosthetic fitting crap first, though."

"How long do you think it will be?"

"I don't know; I guess it takes as long as it takes. I'll have to learn to use it, too. So, more physical therapy. I screwed that up, pushing it too hard. That's how I lost it."

"Whenever you're ready, I'll help you get back into shape for the flight physical. After all, that's what us Jar Heads do best—PT!" he beamed a little as he said it, trying to lighten the mood.

Sam still wanted Ali in the worst way. He never got over her from their flight school fling years before. He offered to help her train, thinking it might spark some renewed interest in him for her.

"That's sweet of you, Sam, but I don't know where this's going for me or how long it'll take. I'll keep your offer in mind, though," she paused and then continued, "Sam, do you know what happened to Michelle? Did she make it all right?"

"I'm not sure where she's at now. All I know is that she was sent to Sigonella shortly after you were sent there. She kept having real severe headaches that the ship's doctor couldn't explain or treat effectively without opiate drugs."

"She asked about you all the time. The story I got from the medical department guys was that the command became suspicious about your and her relationship."

"Yeah, I knew about that. Screw whatever they thought. We were just close friends, for my part. Michelle was always way too attached to me, though. She was Bi anyway, if you know what I mean. I had to remind her a lot that I wasn't into that. She was obsessed with me. Sometimes, it got pretty awkward. Keep all that to yourself, okay? I wouldn't want her to have any problems over it. It doesn't even matter now, I guess."

"Yeah, I'll keep quiet about it. I tell you, though, she didn't like me much. She was kind of hostile towards me." Ali grinned and let out a little giggle at Sam's admission. "Oh shit! That hurt. I can't laugh too hard! Yeah, she knew we had a thing in flight school. I guess she might have been a little jealous and maybe worried about us picking up where we left off since you showed up at the same time as me."

"Hmm, I suppose that explains a lot. I'm still pretty hung up on you, too, even after all this has happened."

"Oh, Sam, get over it. I appreciate your feelings, but I'm not looking for that right now. Shit - look at me, I'm a mess, a cripple! I don't even know where my life's going from here."

Sam was a determined Marine, though. He wouldn't give up that quickly — "adapt, improvise, and overcome," as Marines like to say. He made her an offer he thought she couldn't refuse.

"Tell you what, when you get the okay and get out of here, I'll help you train, get back into shape, and pass that Flight-Physical! I'll make you tougher than a WM drill instructor, girl! How does that sound? I'm here for you, Ali. Just ask—no strings." She let out a loud belly laugh at Sam's offer, then winced a bit more at the pain it caused in the stump of her leg.

"Owe - Damn, that hurts! Okay, you wise-ass Marine, I'll give it some serious thought. Now get out of here; you're making my leg ache with your Marine bullshit! But stay in touch. Thank you for coming to see me, Sammy! You made me feel much better, except for my God-damned leg." Out came another little chortling giggle of delight over that last bit of their conversation as he left.

"Okay then, by your leave, Ma'am! You get some more rest and heal up. And I'll be checking in on you from time to time. I'll also see what I can find out about Michelle for you. How does that sound?" came Sam's response.

"Thanks for that, Sam! Now get out of here, you Jarhead!" she said, letting out another laugh as he left her room.

It was the first time she'd had a good laugh since the accident. She felt better that Sam cared enough to go out of his way, visit her here, and show her some support. It made her feel less isolated and a little better that someone actually cared about her.

He stood by her side through this time, offering encouragement and reminding her of her incredible resilience. With his support, Ali felt she could find herself again and might pull it off—maybe even get back to flying. Getting her wings back seemed less out of reach, with Sam around offering to help her.

Dr. Carver, the Shrink

Several days later, the door to her hospital room swung open slightly. A serious-looking, balding, slightly overweight figure wearing a crisp white lab coat and old-fashioned round-lens glasses stuck his head in her room. Ali's visitor was Dr. Carver, the shrink she'd met and fenced with at Sigonella. It turned out he was one of Balboa's staff psychologists.

"Lieutenant-Commander Cabrillo, I'd like to chat with you if you have the time. May I come in?"

"Well, well, well, look who's here! Sure, I suppose that would be fine. What brings you here, Dr.? Why do you want to chat with me again? Why are you even here?" Ali was very suspicious now, seeing this character again, wanting to do another interview.

"Oh, I'm sorry; I didn't mention I'm on staff here. Or did I?"

"Must have slipped your mind, Dr."

"I'd like to continue our talk about your experiences after your helicopter crash. Particularly after your most recent mishap. Would that be okay? Do you feel like chatting about it at all?"

She was suspicious that this might be one of those interviews to determine her psychological fitness to return to flying for the Navy, since Conway had mentioned it at Bethesda.

"Okay, what do you want to know? We've already had this chat several times. Don't you have all the doctor's comments from Bethesda or Sigonella? After all, it was you who talked to me then."

"Well, yes, I have your records, which are well documented. It's all standard procedure. We want to verify how you are doing since it's been a while, and you've been through quite a bit since then."

She scowled at him and said, "Okay, ask your questions, then. Let's get it over with."

Carver sat across from her, flipping through a bundle of papers he'd been carrying, her records. "Lieutenant-Commander, you've experienced several traumatic events during and after that crash. Can you tell me how you feel about the entire experience now?"

Ali took a deep breath, trying to push down feelings she'd had about the people who were killed, wounded, crippled, and maimed for life as she had, especially those that had been killed. Memories of the crash were fuzzy, but clear enough to haunt her thoughts most nights.

"I'm doing my best to move on, but it's hard to deal with sometimes, especially for those I knew well who didn't make it. I keep replaying that mission in my head, trying to find mistakes I might have made and trying to accept I couldn't have helped getting hit with rocket fire, like we did, no matter what I'd planned for the mission."

Carver nodded understandingly. "It's normal for a leader to have these kinds of feelings after a traumatic event like this. Have you been experiencing any other problems, such as nightmares or anxiety attacks?"

Ali hesitated, unsure if she wanted to admit the full extent of it. She knew she needed to fess up to some extent, but didn't want it to sound like she was having severe coping problems, either.

"I don't experience nightmares. I do have trouble sleeping sometimes, not every night, but sometimes. It's fading as time passes, but it's still there."

Carver listened attentively, jotting down notes as she described her feelings and experiences to him. "It's important you acknowledge these feelings, Lieutenant-Commander. It's okay that you feel this way. You must work through these emotions before expecting to return to duties where you're responsible for others again."

The evaluation continued. Ali opened up more about her guilt at surviving the crash when others didn't. She talked about the pressure to prove herself again and try to lose the guilt of not being better at anticipating things despite her fears. Throughout their conversation, Carver pushed for more, trying to get her to confront what he thought were her genuine internal conflicts, the ones she didn't want to talk about.

"Lieutenant-Commander Cabrillo, your personnel records show you've been on your own since you were quite young. Your father died, and your mother could not cope with raising you herself, so you were sent to live with your grandmother. She passed away a few years later, leaving you on your own. Would you say that these experiences adversely affect you now?"

The line of questioning made her uncomfortable. These questions seemed irrelevant to the crash, the battle, or her feelings about it and made her feel defensive. It seemed to her as if he was probing for weaknesses to use against her.

"Well, I'm not sure how to respond to that, Doctor. How is that relevant to my current situation?"

"I'm looking for you to tell me about your life before the Navy and what might have motivated you to join the service. It must have been very tough to pull yourself up from those conditions, go to college and then go through officer and flight training, and become the successful officer you've been up to this point, all on your own. Tell me about that. What motivated you?"

"Okay, I would say that my primary reason was my uncle Collin. He was a Naval Aviator when I was a young girl. My Dad and I spent many weekends with him when he was not deployed at several duty stations. Besides my father, he was my hero and role model.

I wanted to fly Navy jets like him, and you're right; it was very tough for me after he died. I also had several years where I just drifted after Grandma passed and didn't know what to do or how to get started at anything."

"What was your relationship with your mother like before he passed?"

"She was always quite cold. Even a little distant. I used to think she never really wanted me. She wasn't abusive physically, but she was not interested in me."

"I always felt like I must have been a worthless daughter or something. But truthfully, she just had her own problems, I guess."

"Things just got worse with her after my dad passed. She didn't want to deal with me, or anything else for that matter, at all."

"So, what got you going then? How did you pull out of that?"

"I got a job that helped pay bills, but I wasn't happy or satisfied with it, so I went for it. I signed up for college classes and spent all my time working or attending school. When I graduated, I applied to OCS and got accepted. The rest, as they say, is history."

"Have you noticed any mood or emotional changes since your latest injury and loss of your leg?"

"Seriously, Doctor? Look at me! To answer your question, I'm frustrated and irritated at my situation! How would you feel?"

"Well, yes, I can understand that. Do these feelings cause you to be on edge or nervous?"

"Nervous? No, edgy? I just told you it's very frustrating for me to be in this condition! What else can I say about it?"

"What about feelings of guilt or shame related to the crash? Do you still feel you were the cause or responsible for it?"

"I did at first, but others, like yourself, have pointed out that I couldn't have ever predicted those ISIS rockets blowing my helicopter out of the sky and that it wasn't my fault. Do I feel responsible? Yes, I do."

"People were flying with me that day that I've known for a long time who are dead or in as bad or worse shape than me! That happened when I was in command, so yes, I do."

"Okay, Lieutenant-Commander. I think I have enough for now. I'm sure we will talk again soon. Thank you for allowing me to visit with you this afternoon." He got up and just left her alone there in that hospital room.

Ali didn't think their conversation had helped her case much.

"Shit! This isn't going well. I wonder what this prick is really up to?" Her hands fidgeted with the sheet on the hospital bed while she brooded about it.

Carver returned to his office, wrote his preliminary report on the interview he'd just had with Ali, and contacted Captain Albert Smith, the chief medical officer on the medical review board. Smith had been reviewing Ali's records at the request of a more senior commander, the details of her crash and injuries, and the current status of her recovery.

Smith was a veteran trauma surgeon of the Iraq war. He'd seen many cases of traumatic stress in wounded soldiers. One of his jobs at the Naval hospital was to evaluate patients for retention in the Navy or recommend discharge for medical or psychological reasons.

His office phone rang; Smith answers: "Yes, Yeoman, what is it?" "Captain, you have a call from Dr. Carver."

"Very well, I'll take the call."

"Yes sir," Yeoman Randall patched Carver through. The line clicked, and Smith spoke first.

"Dr. Carver; where are we with Lieutenant-Commander Cabrillo's case?"

"Captain, I've just completed a detailed interview with her. I have the report you asked for ready. May we meet to discuss her case?"

"Yes, I'll have Yeoman Randall set up a meeting with you tomorrow."

Smith punched the button on his office phone for his Yeoman. "Petty Officer Randall, please make an appointment with Dr. Carver to meet with me tomorrow afternoon."

"Yes sir, will do," was her only response.

Carver meets Captain Smith

The following day, at 14:00, Carver arrived at Smith's office on time. When his office phone rang, Smith answered with an authoritarian and demanding voice: "Yes, what is it, Yeoman?"

"Sir, Dr. Carver has arrived for your meeting."

"Very well, send him in."

"Yes, sir."

Yeoman Randall relayed the captain's request to Carver, "Dr. Carver... Captain Smith will see you now; please go in."

"Thank you, Yeoman." Carver opened the door to Smith's office and walked in. "Good Afternoon, Captain."

"It's customary to knock before you just walk in, Carver!" Smith was big on etiquette and protocol. He was always good at correcting others' faults and failures.

"Come in anyway and have a seat. So how did it go with Lieutenant-Command Cabrillo?"

"Well, sir; generally she's in good condition psychologically, but she has some problems with guilt over the loss of people who were under her command when the crash happened."

"I see; go on."

"She's resentful of her present condition. She seems determined to return to flying as soon as possible."

Smith answered him sarcastically, "Yes, I'm aware of her ambitions." "We'll see about that! That's a decision the medical board will make, one way or the other," and, of course, Smith was the ranking officer on the board!

Carver continued, "She also stated she's had some problems sleeping and coping with feelings of responsibility. Survivor's guilt, I would say. She indicated she wasn't having problems with re-experiencing or nightmares. She seemed evasive discussing her current condition and how she felt about it."

"You think she was being evasive, then?"

"I believe she was not being evasive but guarded with things she would open up about. She seemed more concerned about answering questions she felt might be perceived as negative."

"I see, Dr.; I want you to dig deeper. I need to know if there's any possibility that her return to flight duty would be affected by her mental state or her physical condition because of the injuries she's sustained, or otherwise, for something else."

"Understood, Captain. I'll do another interview with her before she's discharged."

"Good, get it done then. That'll be all."

"I'll report back to you after I've interviewed her again with a full written analysis."

"Very well then; thank you for coming by."

The captain was looking for plausible arguments to push Ali out of flying. His experiences in Iraq, treating and dealing with soldiers with catastrophic combat injuries and PTSD, molded his opinion that anyone, especially women who were seriously injured and traumatized in combat or any dangerous situation, was not suited to be returned to situations with the same levels of risk.

In Ali's case, he felt he was protecting her and others like her who might be affected by their command decisions, from exposing them to further trauma. Ultimately, his responsibility was to protect the interests of the Navy and other interested parties, in this case, the Admiral at ComNavAirPac!

He wouldn't take that risk for her or anyone else. As the ranking officer on the medical review board, the other officers would follow his lead and vote with him on their decisions. At least his medical staff would.

Sam Helps Ali

A few days later, Sam stopped by to see Ali again. The door to her room was already half open, so he walked in with a big grin.

"Hey Sam, back again for more abuse, are you?" she quipped.

"Ha, yeah, I guess you might say that. I have some news for you about your friend Michelle."

Now, her interest was piqued, and she asked, "Really? What did you find out?"

"You won't believe this, but she's right here in San Diego. She lives in Old Town, in a house she inherited!"

"What? How did you even find that out?"

"I called around. One of my buds at BUPERS and another guy I know at the VA hospital told me about her and where she is. She was released from the Navy with a medical disability."

"He also said she's not doing well. The crash seems to have scrambled her brain somehow. She kept having severe headaches and memory problems."

"Oh, no! Did you get her cell number or a contact phone number?"

"No, they wouldn't give me that. Honestly, Ali, he wasn't too comfortable giving out her address either, but he was familiar with her case and knew who you were, so he gave it to me."

"I'd pay her a visit for you, but I'm not sure how she would react since she didn't like me on the ship, but at least you know where she is now. Here's her local address." Sam handed her Michelle's hand-scribbled address on a yellow Post-it note.

"Shit! I'll have to visit her when I get out of this place. Oh, I had a visit from one of the house shrinks today."

"Really? How did that go?"

"I don't think well. It was that Doctor Carver from Sigonella and Bethesda! He kept asking me the same stupid questions I'd already been asked. I got a little pissed at him and probably said too much."

"Well, don't worry about it right now. It's more important for you to finish your treatments so I can start helping you train for that physical."

Ali felt this wouldn't happen as he'd promised it would. Her voice took on a grave tone. "Oh, Sam, I hope I can still do it. I'm afraid they're evaluating me now for a discharge. I think I went too far with this, Dr. Carver, today. I guess I'll see how it goes."

"Don't stress so much about it, Ali. If it's possible, we'll get it done." Sam tried to lighten the mood a bit. He could sense her doubt and apprehension. What he'd said seemed to pick up the mood of the conversation for her a little, though.

She grinned at him and quipped back with her sneaky little giggle to amuse herself, hoping it would work out as Sam had said. "Okay, Lieutenant Colonel Jarhead!"

"Ali, I have to run now. I've got squadron business to tend to. I'll stop in again in a few days. Listen, here's my cell number. Call me if anything comes up or you need something."

"Okay, Sammy, thanks, that's sweet of you. Catch you later."

Captain Smith

Captain Smith, the Chief medical officer on the medical review board, dropped by to visit her later that day. He wanted to discuss her future with the Navy. The door to her room was already open. He knocked on the door frame anyway, out of courtesy, and introduced himself. "Lieutenant-Commander Cabrillo, may I come in?"

She took one look at him from the hospital bed and did a double take. He was a captain, wearing khakis with ribbons and eagles on his collar, not hospital scrubs like everyone else.

Her first thought was, "Oh shit, what's this now? This can't be good!" "Why yes, Captain sir, do come in! To what do I owe this visit, sir?" She felt she had to be formal, since she didn't know who he was or what he was there for.

"Right to the point, I see. Well, that's good; I like that in an officer. I thought I'd drop by and discuss your future career aspirations since you will leave us soon."

"I will? That's news to me. No one has told me that, sir. Why would you say that? I'm sure you are aware I'm still undergoing physical therapy."

"Yes, I'm aware of your current condition. However, Dr. Bélanger informs me you're close to completing it and doing well. Based on that, I'd like to discuss your future career plans with the Navy."

"Captain, I intend to regain my flight status and will work to do that, sir."

"That's a commendable attitude, Lieutenant-Commander, but not very realistic, I'm afraid."

"Why do you say that, Captain? I'm one of the best female Aviators the Navy has. My Fit-Reps should reflect that in and of themselves!"

"Yes, indeed they do, but you won't likely be able to handle the physical rigors of flying again. In particular, your back injuries will prevent you from being able to handle hard landings on a carrier deck or any hard landings of any kind, for that matter, ongoing."

He continued: "That doesn't keep you from other types of duty, such as instructor, aviation maintenance officer, or anything else that doesn't involve the physical rigors of flying. The medical review board is not convinced that you won't encounter psychological issues either, given your responses to Dr. Carver's interview recently."

"Doctor Carver, I see. Is that what this is about, Captain? Does Dr. Carver feel that I've got survivor's guilt or some mental problem that might cloud my judgment or something like that, sir?"

"I don't know what his reasoning is, but we both feel that it would not serve the best interests of the Navy to allow you to continue flying."

"Sir, with all due respect, don't I have the right to at least try to pass the flight physical and whatever else I need to do to return to flying?"

"Well, Lieutenant-Commander, your evaluation before the full review board will be in ninety days. You have until then to prepare and convince the board that you're fit and competent for it."

"So that's it then, Captain? I've got just ninety days to prepare even though I'm not finished with physical therapy here yet, sir?"

"I'm sorry, Lieutenant-Commander. That's the best we can do for you. We have regulations and flight standards we must follow as they exist. In the meantime, you can take leave or receive temporary orders to North Island, where you will be assigned to the Admiral's staff."

"You can train while you're there, off duty, and do what you feel is necessary. Do you have any other questions?"

"No, Captain, I don't. It seems that's my only option. Thank you for the opportunity to give it at least a shot."

"Good luck then, Lieutenant-Commander Cabrillo. Goodbye."

Smith turned around and walked out, leaving Ali dumbfounded, frustrated, and shocked as it sank in that this would probably be it for her flying career—her only genuine passion. She would not give up. Thinking about what to do next, she called Sam to tell him what had just gone down and take him up on his offer.

She punched his number into her smartphone and waited for him to answer.

"Lieutenant Colonel Nash."

"Sam, it's Ali. Do you have time to talk?"

"Sure, Ali. What's going on? Are you okay?"

"Not really; I just had a visit from the head of the medical evaluation board! They are giving me 90 days until they meet to decide whether to let me keep flying!"

"What the hell! How can they do that? You're still in physical therapy, aren't you?"

"Officially, yes, but it's close to being done. The therapist just hasn't mentioned it to me yet. He also said I would have orders cut to North Island to work in the interim after they release me."

Sam was a little surprised. "Doing what?"

"I don't know. He didn't tell me that. He only said I would be assigned to the Admiral's staff until the board's finding. After that, I don't know. What do you think, Sam?"

"It sounds like they aren't giving you a choice, Ali. Tell you what, take the orders, and we'll start training you here at Miramar if your therapist gives you the go-ahead. Ask him how hard you can work yourself, though."

Ali was silent for a few seconds. She didn't know what else to say, only that she appreciated his help. "Thank you, Sammy. You're a genuine friend for doing this for me. I won't forget it, Marine!"

"You can thank me later; we've got work to do now. We'll start here at Miramar. There's an excellent sports complex here we can use. They also do fitness assessments, so you'll have a good baseline to argue your case with the medical board when it comes up."

Ali and Michelle

Since Sam had given Ali Michelle's address, she made her way there. Michelle's place was on a hill above the harbor. Ali climbed up a steep flight of stairs to the front door. Stairs were still a little challenging for her and the new leg. It got easier as she worked at it, but it was still tricky. She knocked on the door several times before Michelle answered it. Ali just said, "Hi, Shelly. It's been a long time. How are you doing?"

Michelle was dumb-struck and screamed hysterically, nearly fainting! Ali grabbed her before she fell to the floor and steadied her. Michelle completely broke down seeing Ali standing there in front of her, with part of her leg missing, wearing a prosthesis, but most of all, still alive! "Calm down, Shelly, calm down! I'm all right; I'm here, okay? Come on, let's sit down on your couch here."

Michelle poured out her pain, her eyes two sizes bigger than normal. Pleading and whimpering like some little kid, streaming crocodile tears down her cheeks. "Oh God, Ali, I was so worried about you! I couldn't find out anything about you, where you were, or if you were all right, dead, or what! Oh, my God!"

She curled up in a fetal position beside Ali, leaning on her as if she were her mother. She trembled like she had chills, whimpering and crying her eyes out. Ali wasn't sure whether Michelle was happy, relieved, angry, or something else. The scene Michelle made affected her, too. She started tearing up as well.

"Shelly, I'm sorry I couldn't contact you; I was laid up at Sigonella and Bethesda for a long time. After I lost my leg, they sent me here to the Balboa prosthetic center for this thing." She looked down at it with daggers in her eyes. That prosthetic horror attached to the stump of her left leg, with a combination of pain and anger. She didn't mince words about her hatred of it. She couldn't stand the thing.

After a while, Michelle calmed down and asked about what had happened and how Ali had found her. "So, how did you ever find me? I never gave anyone this address. It was my grandparents' house. I never lived here either before I was retired from the Navy."

"I got it from a friend who got it from the VA. They checked to see if the VA had any information about you they could get for me." She didn't mention to Michelle who she got it from.

Michelle leaned against her a little closer, paused, then spoke softly, just a whisper, like a little girl talking to her mother after a long separation… "I don't care how you got it. I'm happy to know you're all right, finally."

"I didn't say I was all right, Shel. I mean, look at me." She looked down again, staring in disgust at her leg.

"What happened? Couldn't they save it?"

"They tried, but I screwed up - I overworked it too much, too early. All the surgeries to it just failed, and I got a bad infection from the bone grafts. So, they had to take it. They told me I probably would have lost it anyway because of the problems with the repair work and infection."

"What do you mean, you overworked it?"

"I want to get back to flying, Shelly. I was trying to get back into shape for the flight physical. I spent so much time lying on my back in hospitals that I was a flabby lump in terrible physical condition. Even after I could walk again, it was exhausting just walking around for short distances, let alone doing any serious physical training. So, I just overcompensated. I worked everything too hard."

Michelle was incredulous, scolding her: "Ali, what the hell! What were you thinking? After all of this, why would you want to go back to flying, risking your life again? For what? Look at yourself, girl!"

Ali's response was simple and direct. One Michelle couldn't fathom. "It's all I ever wanted to do, Shel. It's pretty much everything for me."

"I guess I can understand that. It was never that big a thing for me. I enjoyed flying, but I hated all the military bullshit all the time. At first, when they told me I was out, yes, I was hurt and insulted.

They just threw me out like a piece of used-up junk. I wasn't really up to it anymore, anyway. I still have lots of problems from that crash."

Ali wanted to know, so she had to ask, "Really, what kinds of problems, Shel?"

"I still have terrible headaches and memory problems. Concentrating on things is hard for me, too. Sometimes, I lose my balance and fall for no reason." My monthly cycles are screwed up, too. Sometimes, I don't even have them at all. I can't get the crash stuff out of my head, either."

"What do you mean? What is it you can't forget about, Shel?"

"Seeing those guys getting torn apart by the rifle fire and explosions going off all around us! Especially that SEAL Master Chief and his guys, and the smell, Ali, the stink. I can't get it out of my head."

Ali looked at her wide-eyed. Stunned by her description of the action during the ground fight! "What do you mean the smell, Shelly?"

"You know... that gunpowder smell from the rifle fire and explosions, the acrid burned chemicals stink, mixed with sweat and adrenaline! That metallic taste in your mouth, the stink of blood and shit in the air, the dust on the wind. You didn't have any of that?"

"I really can't remember any of that, Shel. I was pretty messed up, though. They shot me up with morphine right away."

"You're lucky; I can't get it out of my head!" Michelle shivered and wept a little, opening up to Ali about it. She continued pouring it out.

"They told me I might have some kind of traumatic brain injury thing, TBI for short, but they couldn't find anything physical or do anything about it except give me drugs. Now, the VA won't renew my pain scripts. They told me the drugs were too addicting, but nothing else worked. I didn't know what to do then, Ali."

"I'll do what I can to help Shel, but I'm trying to pass this medical board review, so I have little free time right now."

"How are you even going to do that, Ali? Where are you staying? You're out of Balboa now, aren't you?"

"Sort of; I'm in outpatient care. I have to go in for checkups like weekly so they can look at my leg and this fake one I have to wear. They cut me temporary orders to North Island to the Admiral's staff at AirPac. If I pass my evaluation and the medical review board approves, I'll go back to flying and try for another squadron. If I don't, I'll be washed out, like you were."

"Where are you staying? You could stay here with me for a while!"

Michelle looked at her with big, puffy, sad eyes and a pained expression, practically begging Ali to stay with her.

Ali was sympathetic but firm; "Oh, Shel, I can't do that, but I'll come by and visit you as much as possible. I have to concentrate on working out and getting in shape. Plus - I don't know yet what I'll be doing at AirPac either."

Disappointed by her answer, Michelle replies, "Okay. You going to train at North Island?"

"Yes, and at Miramar."

"What, Miramar, why there? It's a Marine base!"

"Yeah, it is, but I have some people there who will help me. They've got an excellent sports complex there, too, and they also do fitness evaluations."

"Oh, I see." Michelle suspected she knew whom Ali was talking about. They sat and caught up for several more hours until Ali left.

"Well, Shelly, I have to go now, but I'll be in touch, I promise. I wrote my phone number down on this piece of notepaper for you. If I can help or you want to talk, you can call me whenever. It's probably not a good idea during work hours, though. I don't know what that's going to look like yet."

"Okay, Ali, I'm so glad you came by. I've missed you so much; you don't know."

"Sure, Shel, I've missed you too. We'll be chatting again soon. Stay safe. Be careful taking those drugs!"

Ali walked out the door, down the stairs to her rental car, and drove off to North Island to check in with ComNavAirPac, her new command, and get a billet assignment for the BOQ, the 'bachelor officer's quarters' as they're called.

Chapter 9: Turning Point

Ali Goes for It

After checking in with AirPac (Commander Naval Air Force Pacific), Ali headed for the single officer quarters, where she would stay while working there, waiting for the medical review board's decision. "Stay or go. Their decision," she thought to herself.

"Screw Smith and the rest! I'm not giving up that easy."

She'd already done everything possible to return to flying. After today's events and her emotional reunion with Michelle, she needed to get some rest, crash early, and review her plans to prepare for the medical review board's evaluation.

Musing over the term, she thought a bit more about it, "Hm, 'crash': not a very good term for a pilot! I'll just call it 'resting' for a while instead."

She grinned a little, amused by the thought. It seemed like a funny contradiction of terms, considering a crash was why she was here. Ali slept through the night for ten hours until the following day. A wake-up call buzzed her phone at 06:00.

"Damn, what happened? It's tomorrow already, shit! Wasn't expecting that!"

Ali got up, put herself together, and headed for her new office. Her temporary assignment was 'Staff supply Aviation Supply liaison officer' to fleet air support for aircraft maintenance for all Pacific fleet helicopter squadrons and the various parts vendors supplying repair parts not otherwise available.

An 'expeditor' job, a do-nothing desk driver, just talking on the phone, going to meetings with other staff officers, and reviewing supply and aircraft maintenance reports that enlisted staff did all the work preparing for her.

She hated it: "This just isn't my sandbox—flying a desk, answering phones, sitting through boring meetings all day!"

She could see aircraft and helicopters coming and going from her office, taking off and landing. She stewed about it inside: "This gig is just God-damned torture!" Watching them come and go, take off and land, a mix of longing and determination ate at her.

"This fucking prosthetic leg! I can't stand it! I've got to get back to flying!" As far as she was concerned, it was an unpleasant reminder of the day everything went sideways.

It was time to get to it and get back in shape. She punched Sam's number in her office phone and got the show on the road; the clock was ticking. His phone rang in his squadron's office at Miramar: "Lieutenant Colonel Nash's office. This is a non-secure line; may I help you?"

She was surprised he didn't answer himself. She hadn't thought about him having a personal office staff.

"Yes, Lieutenant-Commander Cabrillo at ComNavAirPac, for Lieutenant Colonel Nash. May I speak with him, please?"

"I'm sorry, Ma'am. He's not in his office now. Shall I send him a message?"

"Yes, please let him know I called and have settled in at my new command, and I'm ready to start on the project we discussed."

"Yes, Ma'am, I'll give him the message."

"Who am I speaking with, please?"

"Staff Sergeant Nguyen, the squadron admin NCO."

"Okay, thank you, Staff Sergeant". She wondered why he wasn't in his office. Whatever, he was a pilot too, maybe out flying or off at some meeting somewhere with who-knows-who or what.

"Well then, I'll just get started here myself."

That evening after work, she headed to the base gym and started with the routine Dr. Bélanger had given her to follow. She worked hard. That hard-driving personality started coming back and taking over, the one that caused the failures with her ankle and foot before she lost them.

A couple of days later, sitting in her office working on some maintenance paperwork for the NARF, she got a call back from Sam. Her smartphone buzzed like an angry bee.

She picked up, "Lieutenant-Commander Cabrillo."

"Hey, Ali, it's Sam!"

"Oh, hey, Marine, what happened to you? I thought you abandoned me!"

"What? No way! I was called off to meetings with the Air Group's commanding general for a few days. I didn't forget about you, but I'll be off doing carrier quals for a few weeks."

Ali didn't like the sound of that. "So, you won't be able to work with me then?" she asked suspiciously.

"Not till I return, but I've got somebody here that will."

"Really? Who might that be?"

" Gunnery Sargeant Barrett. She was a drill instructor before she was assigned to my squadron. I told her about you and what happened.

She volunteered to help you out while I'm away."

"Oh really? A real hard ass, no doubt."

"Ali, if anyone can get you toughened up enough to pass that physical, it's Gunny Barrett."

"Okay, so how will that work, then?"

"I'll hook you up with her. The two of you can work out a schedule; how does that sound?"

"Sounds fine. Thanks for the effort, Sam?"

"Okay, I'll give her your contact info. She'll call you tomorrow afternoon."

"That sounds great, then. Thank you again, Sam. I appreciate it very much!"

"Hey, I'll do anything to help you out, Ali. I'm sorry, I can't do it myself just now. Gunny Barrett will get you going."

The next afternoon, Ali and Gunny 'Susan' Barrett started things rolling after a brief phone conversation. As Sam promised, Barrett called her: "Lieutenant-Commander Cabrillo, how may I help you?"

"Hello, Ma'am; I'm Gunnery Sergeant Barrett at Miramar. I'm calling to see when you might want to start your physical conditioning and training with me."

"Oh, hey, Gunny, I damned sure am! How do you want to go at this?"

"Ma'am, it would be best if you could come to Miramar. I can work with you here after hours. We have a very extensive training facility and sports complex here. Everything you'll need."

"Okay, I can be there around 17:00. How's that sound?"

"That's fine, Ma'am, I'll meet you at the Sports Complex. Just call me when you arrive."

Ali took off for Miramar's Sports Complex to meet with her after work that day. Barrett was already waiting for her at the front door when she arrived. Phone calls or text messages weren't needed.

Barrett recognized her as soon as she got out of her rental car. A female Navy officer in uniform with a prosthetic leg sticks out like a sore thumb on a Marine base.

Barrett waved her to the front door of the sports complex. "Afternoon, Ma'am."

"Afternoon, Gunny. Just call me Ali while we're working together, okay?"

"Sure, I can do that, Ali. My first name is Susan. If anyone else is around us, though, call me Gunny. We're pretty strict on protocol around here, so if there's anyone else around, Gunny's fine."

"Okay, sure, that's fine, Susan."

Ali worked with Gunny Barrett for a few weeks and started feeling a newfound sense of purpose—getting that old jazz back. She threw herself into training workouts and rehabilitation. Her prosthesis started symbolizing resilience and determination rather than limitation.

Now, if she could convince the skeptics and the higher-ups on the medical board, she might even pull this off! Facing all the skepticism and bureaucratic bullshit wasn't easy, but she had the drive and the determination.

The physical demands were intense, pushing her to the limits of her endurance. Gunny Barrett wasn't easy on her, either. She pushed Ali hard.

"Is that all you got, Ali? Do you want to get back in the cockpit or not? Get your ass moving! Give me another five laps! Move it, Sailor!"

Push-ups, sit-ups, running the obstacle course behind the gym, pumping weights. Whatever it took, she did it! "Keep it up, sailor. Five more reps, add more weight, move it!"

Barrett was relentless, pushing her like a green boot in basic training at Paris Island.

Ali welcomed the challenge with open arms, fueled by determination. Every day, she pushed herself further and harder, determined to prove she was still capable. It didn't take her long before her hard work paid off big time.

Her strength and agility returned, surpassing even her expectations. She was more like her old self again!

Ali knew in the back of her mind that this training was only half the battle. There were lingering doubts. She knew there would still be questions about her psychological fitness after the crash and all the trauma, pain, and guilt she carried around that she wasn't even aware of yet.

The real question for her was how to show that she could deal with that kind of stress in the future, mainly to Smith's satisfaction. She could read him like an open book. She suspected he was the prick who wanted her to quit! She needed to show him she could conquer the lingering mental barriers and be an effective Aviator again. She didn't know Smith had already thought of that…

Ali was soon in the final phases of training, approaching the end of her 90 days before the board's evaluation. She had already completed all the technical competency reviews and medical tests, passed the rehabilitation program with excellent evaluations, and completed all physical fitness testing.

So far, she'd been cleared of any other health concerns. She was otherwise fully functional but still struggled a little with running until she 'acquired' a running blade. Wearing it, she could run nearly as well as before the accident. No more stiff-legged limp for her!

Sam and Gunny Barrett were duly impressed with her progress and gave her an enthusiastic thumbs-up. Sam especially:

"You are damned impressive, Ali. I didn't think you could pull this off, but here you are!

"Thanks, Sam. And you, of course, Gunny, I couldn't have done this without your help."

"My pleasure, Lieutenant-Commander Ma'am! Always enjoy screwing with sailors, anyway!"

They all three laughed out loud at that one. There was still the matter of the psych evaluations, though.

Dr. Carver: Round Three

Ali was notified 'Officially' through her command at North Island that she was ordered to meet with Dr. Carver 'again' for another follow-up interview 'code-speak for a psych evaluation.' He didn't contact her directly by phone or email but notified the Admiral's chief of staff to direct her to meet with him. It was a hand-delivered official memo from the ComNavAirPac Chief of staff's office:

"From: Commander Naval Air Force Pacific Fleet

To: Lieutenant-Commander A. L. Cabrillo.

Subj: Psychological Evaluation.

You are directed to report to Dr. Charles Carver at Balboa Hospital for further psychological evaluation as soon as possible before your evaluation by the hospital's medical review board and ruling regarding your request for reinstatement to Aviation duties. Please arrange an appointment with Dr. Carver's office ASAP."

"Well, this is damned suspicious," reasoning it out in her mind:

"Why would they send this memo through the chief of staff's office instead of just notifying me directly? And why was that geek analyst Carver still in on it?"

Doubts soon turned into dark suspicions. The possibility of private discussions between Smith and her command, or even worse, the Admiral himself, sent a cold chill through her.

"Smith must've been whispering in the Chief of Staff or the Admiral's ear all this time! Damn it! Or maybe they planned it like this from the beginning. This is not good!"

Now, she was even more nervous about being railroaded out of flying. She was really at a loss for what to do. She thought talking to Sam might give her a better perspective:

"I'll just call Sam and see what he thinks about it." She pulled his office number from the contact list on her smartphone and waited for him to pick up as it rang.

"Lt Colonel Nash, this is not a secure line; how may I help you, sir!" came his greeting. "Hey Sam, it's Ali. Do you have some time to talk?"

"Sure, Ali, what's up?"

"Can we meet somewhere? I have what I think is a serious development with the medical review board. I need to talk it over with you if you have time."

"Okay, how about the club at nineteen hundred?"

"Great, see you then, Marine."

A few hours later, they met at the venerated 'I Bar' at North Island, an aviator's hangout for many decades. Ali had been waiting for him for a while. He spotted her as he walked in, sitting at the bar by herself. The place was not very busy anyway—it never is when the carriers are out.

"Hey Ali, you're looking pretty unhappy. What's going on?"

"I'm not sure, Sam, but it smells like I'm being set up—like I have been from the start."

"What? What are you talking about?"

"I got a memo today from the Admiral's chief of staff to report for another psych evaluation before the medical review board meeting! The Chief of Staff! They didn't even have the courtesy to route it directly to me! What the fuck! Why would they do that, Sam?"

"You're right. That seems damned strange to me, too!"

"What should I do?"

"What can you do? Show up and go through the moves. You know that's your only play."

"I know, it just smells like that captain at Balboa has been talking to the Admiral or someone higher up all along! Seems like they never were going to give me a chance!" Ali was getting pretty upset at this point.

"Calm down, Ali. You don't know that for sure. I admit it sounds suspicious, but you can't be certain. You know you can't do anything about it even if it were the case. If it's a command decision, that's what it is."

"You're right, of course. Stop being so fucking logical! It just really pisses me off they would do that to me after all I've been through trying to get things back on track!"

"Just go through the moves and see what develops, Ali. Even if they wash you out, it won't be the end of the world."

She shot him a side glance that could've killed for that remark.

"Shit! What do you mean, Sam? I don't know how to do anything else! I don't even want to do anything else! Makes me sick to my stomach to think about it."

"Let's have a couple of drinks and relax for a while, Ali. How does that sound?"

"Ordinarily, that would sound fine. Now, not so much."

She made the appointment for later that week, the following day. When the time came for the interview, she made her way to his office waiting room and pressed the buzzer next to his office door. Ali sat there feeling nervous, frustrated, even a little irritated at the thought of having to do this again with this sleazy little weasel of a psychotherapist.

She knew it would be critical to the board's decision. After losing her leg, going through all the hell of physical therapy, relearning how to walk again on a prosthetic leg, and then even learning how to run on one, she knew it had come down to this. She'd fought so hard to recover, both physically and mentally. Now, with her prosthetic limb fitting seamlessly and her determination even stronger, she had to face this shrink's shit show - proving herself fit for duty once again to him!

Carver opened his office door and greeted her cheerily: "Lieutenant-Commander Cabrillo, come right in. I'm pleased to see you again."

She replied curtly but formally,

"And you as well, Dr." Carver looked at her over his glasses. He was calm and clinical. Ali felt the weight of his scrutiny as he droned on with his line of rehashed B.S. questions.

"Okay, let's review everything that's happened again," he said in a measured, monotone voice.

"Can you walk me through everything that's happened again, please?"

"Well, okay, fine, if I must." She took a deep breath and recounted everything she remembered. Carver listened intently, jotting down notes on one of those yellow office notepads as she answered.

Then he asked, "How did you feel after the accident?"

"Seriously, Doctor? I was bleeding heavily, barely conscious, and my back was broken! I couldn't even feel my legs! That was a good thing at that point, too, since it was crushed from above the ankle and would have been excruciatingly painful otherwise!"

"Yes, I understand the injuries themselves, but I want to get a feel for your state of mind at that point."

Based on this question, she was sure he was fishing, so she was cautious, giving straight answers but trying not to sound disturbed. Ali thought this might help her case; well, maybe it would.

"Okay, I was dazed after the crash. The gunfire and rockets going off around me were very confusing. I couldn't move myself to get out of the helicopter. I also saw a lot of the guys and the ISIS fighters shooting at each other. Some were getting pretty shot up and killed in front of me."

"What happened next?"

"The Corpsman and one of the Seal team guys started trying to get Lieutenant Robbins and me out of the wreck. I was pinned in and couldn't move. Lieutenant Robbins was quite hysterical at that point, so they gave her a shot of morphine, I think, to calm her down. After they finished with her, they tried to get me out and put a tourniquet on my upper leg so I wouldn't bleed out."

"You seem to remember a great deal of it for someone who was dazed and near unconscious, Lieutenant-Commander; how's that possible?" She realized this was more like an interrogation than a psych analysis.

She continued answering cautiously: "I don't remember all of it. After we were evacuated, the Corpsman briefed me on the ship and the Medevac aircraft."

"So, you aren't certain that's what happened then?"

"Was that a question, Doctor?"

"No, just a simple observation—please continue. What happened next?"

"After the Corpsman and Petty Officer Muldoon put a tourniquet on my leg, they gave me an MP5 in case ISIS got too close and made sure I had access to my sidearm until they could get me out."

"Why would they do that if you already had a submachine gun?" She thought that was a stupid question, but kept it to herself.

"So I wouldn't get captured and tortured to death by ISIS!"

"Elaborate, please; why a sidearm and an MP5?"

"The pistol was for me to decide if I wanted to be captured and tortured to death or raped and then tortured and killed. It was for me to decide if I wanted to shoot myself, Dr. Carver!"

"I see. I'm sorry if I sound so probing, Lieutenant-Commander, but it's necessary, and you seem to remember much of it."

"That part I remember clearly, Doctor."

"Why would you say that, given your circumstances and condition?"

"Well, to be perfectly honest with you, Doctor, there are some things a girl just remembers vividly. Like maybe fearing being raped to death or tortured and killed by ISIS. It sticks with you. you know what I mean."

"Yes, I understand. Did you consider this later, after you'd recovered enough to realize everything that happened?"

Carver studied her intently, searching for signs and telltale reactions, weaknesses in her answers, pauses in her speech, or contradictions—anything that might tell him she was still traumatized or outright lying.

"And how do you feel now about it, Lieutenant-Commander? Do you feel confident in returning to active duty flying?"

Ali answered with certainty: "Absolutely, Doctor. I've worked hard to regain my strength and pass the physical evaluations and all the other assessment testing. I'm as good now as I've ever been. I've proven myself in simulations and training exercises. I'm ready to get back in the cockpit and fly again."

Carver gave her the usual speech: "Yes, I'm sure you do, but as you surely know, the Navy has a responsibility to ensure the mental well-being of its personnel, especially given the traumatic nature of what happened to you. How do you cope with the idea, the 'fear' if you will, that this might happen again during any mission in a hot combat area?"

"Doctor, I understand and accept the risks, just like all military personnel do."

Carver paused, briefly jotting down a few notes, and continued with his line of questioning: "Tell me, Lieutenant-Commander, have you experienced any recurring nightmares or flashbacks related to the accident or your experiences after the crash?"

Ali answered each question calmly and honestly, hoping it would be enough to prove her case beyond any doubt. "Doctor, I've spent most of the last two years working all this out. Frankly, it's been hell, but I've stuck it out to get to this point. I have no bad dreams. I've never had flashbacks from it, nor have I spent any time stressing over fears of it happening again."

"If I'm afraid of anything at all at this point in my life, it's that I won't be allowed to go back to flying! I don't smoke or use, and I seldom take a drink. I don't want to do anything else. I don't even know how to do anything else. Don't take that from me!" Ali's last statement was more of a plea than a request.

The interview dragged on and on. She couldn't shake the feeling that Carver was searching for reasons to fail her. Finally, after several hours of what she thought were pointless and inane questions that he seemed to keep asking repeatedly, sometime later, Carver put down his pen and regarded her thoughtfully for a moment. Then he spoke.

"Thank you, Lieutenant-Commander. That will be all. I'll review your case and what you've told me here and report everything I have to the review board." That was it. The room fell into a heavy silence, the weight of the impending decision hanging in the air. It was all up to Smith and the other review board officers now.

Ali got up to leave, turned, and addressed Carver for what she hoped would be the last time.

"Well, Dr. Carver. I'll take my leave then, and please take seriously what I've just said. I hope you will at least be fair in assessing me."

"Thank you for dropping by, Lieutenant-Commander," was all he said.

She didn't care for his dismissiveness; he wasn't pleasant or cordial at all. He seemed almost robotic—neither friendly nor unpleasant. It gave her an awful feeling like he'd already decided but was going through the motions, gathering information about her to support a decision to wash her out that had already been made.

The Medical Review Board

Three months had passed quickly. Ali's training and evaluation period ended. Then, the Medical Review Board convened to review her case and decide whether to approve or deny her request.

Four commanders were present: two from the hospital's medical staff and two from the ComNavAirPac Admirals' staff, her temporary command. Captain Smith presided as chairman.

Smith opened the meeting: "This Medical board of review will now come to order. We're here to consider and make a final determination as to Lieutenant-Commander Cabrillo's request to be reinstated to flight duties. Commander Stevens, please brief the board members on her current medical status."

"Yessir Captain, the Lieutenant-Commander, has undergone a comprehensive medical evaluation to assess her overall health and physical condition, the specific impact of her limb loss, the use of the prosthesis, and her overall ability to perform essential flight duties."

"The findings are as follows:"

Point number one, her medical assessment: She's completed a comprehensive rehabilitation program, demonstrating her ability to adapt satisfactorily to the prosthesis. She's also been able to return to a high level of physical fitness and maintain it under Navy medical regulations and Aviation Physiology standards."

"Point Number Two: Before she was fitted for it, the hospital's prosthetics center evaluated her specific type of prosthesis and certified it as suitable for aircraft use. Its functionality and adaptability meet all technical standards."

"At the time of its testing, there were no indications it would impede a pilot's performance or abilities in operating most aircraft; however, it may not be suitable for some aircraft types. If that situation arises, it must be reviewed on a case-by-case basis before certifying her to fly those specific aircraft types."

"Point Number Three: Her psychological profile does not indicate the crash, the combat episode she's experienced, or losing her lower leg has caused her severe or ongoing traumatic stress. She expresses guilt for the personnel losses but does not appear to be severely affected by it."

"She does display symptoms of arousal and angry mood swings, but it's not clear if that's related to the incident in question or if it's just part of her overall nature. She's well adjusted, considering her experiences."

Smith responded curtly: "Dr. Carver disagrees with you. He believes she is overcompensating for her dysfunctional early family life and trying too hard to prove to herself she's good enough. He feels this may cause her to make decisions that impede her better judgment, especially in high-stress situations during future combat missions."

Stevens offered a cautious but accurate rebuttal to Smith: "Captain, if I may, those conditions would have been assessed early in her career and are irrelevant now."

Smith wasn't interested in his response: "Duly noted, Commander! However, Doctor Carver's analysis indicated she's carrying significant guilt for losing one of her air-crew and several members of the Seal team and Marines she was carrying on that mission. He feels she may be overly cautious in the future and hesitate when it might be necessary to make life-and-death decisions with the lives of others."

Commander Todd, one of the AirPac staff officers, chimed in: "Captain Smith, I'd like to offer the consensus of the Admiral and those of us on his staff regarding Lieutenant-Commander Cabrillo's fitness to return to flight duties for the record."

"Very well then, Commander, what's the opinion of ComNavAirPac staff?"

"Sir, the consensus within AirPac is she's an outstanding officer and leader, and we'd hate to lose her. The only issues the admiral and the rest of us have are the effectiveness of her treatments or additional therapies she might need in the future, particularly during deployments, and how it would affect her own personal readiness. It could be a point of concern if she needs specialized counseling or drugs, or medical treatment that isn't readily available in forward deployment situations."

Smith interjected, "Indeed, that was one of my reservations about approving her request. Based on the Navy's needs, it would be unwise to approve this request."

All of them, save for one, kept quiet at Smith's last comment. Commander Stevens spoke:

"Sir, with all due respect, she has passed all the requirements satisfactorily. For us to deny her would be difficult for her personally and frankly, an unnecessary loss to Naval Aviation!"

"Again, Commander, duly noted!" Smith was not moved. and a little irritated at Stevens for expressing a dissenting opinion to everyone else. Stevens knew when to keep quiet, so he said nothing further. Smith called for a vote: "I believe we've heard enough; please vote your decisions, gentlemen."

The response wasn't unexpected, given Commander Stevens' disagreement with the captain. It was a 3 to 2 vote against her being reinstated to flight duties. At that point, Smith called his Yeoman assistant:

"Yeoman Finch, please notify Lieutenant-Commander Cabrillo that I want to see her as soon as possible."

"Yes, sir, right away, sir!"

Bad news travels fast!

Ali hadn't been present and wasn't invited to the board's discussion. She knew Smith didn't want or need her there, anyway. Her office phone rang. "Yeah, I'll bet I know who this is."

The phone call sent a woebegone feeling through her.

"Lieutenant-Commander Cabrillo, this is a non-secure line. May I help you?"

"Lieutenant-Commander Cabrillo, ma'am, this is Yeoman Finch at Captain Smith's office. He wants to see you as soon as possible."

"Understood. Tell him I'll be there in a couple of hours."

"Yes, Ma'am, will do."

A sinking feeling overcame her now. Her intuition told her it wouldn't be what she wanted to hear. She left her office on the base at North Island, headed for Smith's office. She walked into the outer foyer of his office suite, thinking to herself, "Moment of truth- this is it."

"Yeoman, please let Captain Smith know I'm her."

"Yes, Ma'am, he's been waiting for you." Yeoman Finch announced her:

"Captain Sir, Lieutenant-Commander Cabrillo just arrived."

"Good. Send her in."

"Yes, Sir, right away."

Ali didn't like this guy at all. She could tell he was on a real power trip. She thought he was rude and abrupt even with his Yeoman.

"Is he always so cordial, Yeoman?"

"Oh, Ma'am, he's in a good mood today!"

"I can tell! Maybe a little too good, I think, for my liking. I'll bet I know why!"

"Yes, Ma'am, I understand."

She knocked on his door and waited for the customary invitation to come in. "ENTER." His response seemed gruff, more of an order than an invitation. She went in feeling apprehensive about what likely would come next.

"Come in, please sit down, Lieutenant-Commander."

"Yes, sir, thank you for the invitation."

"Well, I'll get right to it then. As you know, the board met and voted on your request today. The vote went 3 to 2 against allowing you to resume your return to active flight duty."

Already discouraged and angry, she asked him point blank: "And may I ask Sir what the reasoning for denying my request?"

"It came down to a question of readiness during deployments and possibly dangerous future missions and combat situations. The question about how you might handle future traumatic stress episodes came up and was a factor in the decision as well."

"So, Captain, it sounds like you feel I wouldn't be able to handle it in the future even though men like Master Chief Brooks, one of our Seal team's senior petty officers I was deployed with, had traumatic and repeated instances of stress. Yet he was always cleared after every incident to endure even more. Why was that, Captain Sir?"

"Lieutenant-Commander, that isn't the same thing."

"Sir, with all due respect, I think it's exactly the same! I think you, in reality, don't think women can handle this kind of duty! Do you, Captain? I came here expecting this. When I first met you, you made it crystal clear you opposed allowing me to go back to flying!"

Ali was really pissed now, trying hard to keep her composure, and continued: "I want you to know, Captain, that I'm already preparing my resignation request. I don't see myself flying a desk for the rest of my career!"

"Well, that's your prerogative, Lieutenant-Commander. The AirPac staff officers expressed their desire to see you continue in some capacity. I must be honest with you, though; your attitude is insubordinate. Frankly, I wouldn't want you in my command if, for no other reason than that!"

After that remark, Ali let him have it with both barrels: "Well, frankly, Captain, with all due respect, I wouldn't want to work with you either! You don't seem to respect women's contributions to the service or even your own command! Or maybe you dislike me personally! I don't know. Are we done here, sir?"

"Indeed, we are Lieutenant-Commander! Get out of my sight!" Ali stomped out of his office, ready to explode. Yeoman Finch had overheard their heated conversation, grinned at Ali, and said: "Bravo, Lieutenant-Commander! He deserved that."

"Thank you, Yeoman; if I were you, I'd put in for a transfer!" Finch's remark calmed her down a bit and made her feel a little better. Later, back at her office, the phone rang: "Lieutenant-Commander Cabrillo, this is a non-secure line. How may I help you?"

"Lieutenant-Commander, this is Commander Stevens. Do you have a few minutes to chat somewhere privately today?" His call seemed suspicious, and she wasn't in any mood for any more bullshit today.

"Why, sir, what do you want to discuss?"

"It's about the review board's vote and the decision regarding your case. Do you have time to meet me somewhere? There are facts you need to be aware of."

"What sort of facts, Commander?" Now she was curious.

"I need to talk with you privately about it. Is there anywhere we could meet later?"

"Well, I suppose we could meet at the club here at North Island, the I-Bar."

"Ah yes, that Aviator hangout. Shall we say 1800 today?"

"That would be fine, Commander."

Ali was already stressed, feeling betrayed and let down by the captain's treatment, but her curiosity had been piqued.

She wondered what Stevens might tell her for the rest of the afternoon. "What the hell could this be about? What does he even want to say?"

She headed for the club after work and hung out until Stevens showed up. She didn't know him, but he knew who she was and recognized her right away with that leg.

"Well, Lieutenant-Commander, so nice to see you again."

"Really? I don't recall ever meeting you before, Commander."

"We haven't met formally, but I've seen you and been part of your case at Balboa since you checked in with AirPac."

"I see. So what did you want to tell me then?"

"Well, I just wanted to make you aware the board's vote wasn't the same as Captain Smith's final decision."

"What do you mean? Why would it be any different from what the board voted?"

"Captain Smith's the ranking officer. His decision's final."

Now she was getting worked up and angry all over again: "What the fuck! Why even have this charade, then? Why did he play this little game?"

"It's protocol, standard procedure. He has to do it to legitimize his decisions."

"So why tell me this now, Commander?"

"Because I believe you can fight this if you want to. If the command supports you enough to back you up in an inquiry, the vote was 3 to 2 in your favor! "Captain Smith, Commander Collier, and another one of his staff officers cast the dissenting votes against approving your request."

Stevens made his point. "If there were to be an inquiry, each review board member would have to testify under oath as to their decision and justify it. I have to warn you, though, that even if that were to happen, the chances of Smith's decision being overturned would be slim. He has the final say about it."

"So why would I make a formal complaint and try overturning a senior medical officer's decision, Commander?"

"I can't answer that for you, Ms. Cabrillo; I just wanted you to know the truth. That decision would be entirely up to you."

There wasn't a lot of chit-chat after that. They both finished their drinks. Stevens ended the conversation and said his goodbyes, "Well, Lieutenant-Commander, good luck with whatever you decide to do, goodbye," and left.

Despondent and angry, Ali left the club and headed back to her BOQ rooms, drank some more, and cried herself to sleep that night.

Hail and Farewell

After a day or so of letting it sink in, it was time to turn in her resignation request. She'd just been putting the finishing touches on her 4510 paperwork, making sure everything was accurate, when the office phone rang. "Lieutenant-Commander Cabrillo, this is a non-secure line. May I help you?"

"Ms. Cabrillo, Admiral Callahan. Please come to my office."

'ComNavAirPac' himself was on the line. She wondered, "What the hell? What did I do now? Shit! I'll bet that fucking Smith called him and reported me for arguing with him yesterday!"

She responded, "Yes sir, right away!" Ali had never met the Admiral before. She was a little nervous, considering what she'd said to Smith.

She knocked on the Admiral's door. "Enter," came the response.

"You wanted to see me, Admiral," she answered nervously.

"Yes, Lieutenant-Commander. Please come in and take a seat."

"I understand that your medical board review denied your request to be reinstated to flight status."

"That's right, Admiral."

Callahan's tone becomes sterner and more commanding. "The Chief Medical examiner, Captain Smith, tells me you're unhappy about that. Is that right?"

Ali felt like she had to defend herself against his accusatory tone. "That's correct as well, Admiral. Sir, I've passed all the medical and physical requirements and have a four-O record. I'm also told that this prosthesis has been certified for aircraft use. My feeling is the decision was unfair and biased."

He scowled a little and lectured her on her attitude; "Lieutenant-Commander, the board's decision is not about what you want or suspect. It's about the needs of the Navy and your fitness to perform this type of duty. I understand you may also consider resigning your commission. Is that true?"

"Indeed, it is Admiral. I'm just not interested in doing paperwork, sitting at a desk, or teaching for the rest of my Navy career."

"Well, Lieutenant-Commander," his tone was more conciliatory now, "I would ask that you reconsider that. There certainly must be something you would still be interested in doing for the Navy. I wouldn't want to lose an outstanding officer like yourself over this nonsense."

"You surely realize that flying combat aircraft and helicopters has a half-life anyway, and you would eventually be promoted and moved into a more command-oriented role. You've already got 16 years in. At least think about that. Maybe a move into the reserves for a while to consider your options before you make a final decision."

She hadn't looked at it that way before. Now she had to rethink her feelings about it. "Thank you for your confidence in me, Admiral Callahan, but I worked hard to get where I was before this happened. I feel like I'm being denied for no good reason! I'll consider it in any case."

"Very well then, that's all I ask. Talk with the officer's career counselor if you want more specific options. That'll be all Lieutenant-Commander."

"Yes, sir, thank you for your time and constructive opinions, Admiral."

Back in her office, she was unsure of her next move. Weighing the pros and cons seriously, she had to ask herself:

"Should I just throw it all away? I only have four years left before I can retire with a full pension and a disability rating. But what else would I even do if I stayed in? Maybe Sam has some ideas. Maybe he can give me his perspective."

She called him, this time on his private cell number. "Nash here," he answered, sounding busy and frazzled.

"Hey Sam, it's Ali. Are you busy? Do you have a few minutes?"

"Not at the moment. What's up?"

"I need to talk with you."

"I can't right now. Can we meet at the club tonight?"

"Sure, that would be fine. 1800 work for you?"

"Okay, see you then."

"Okay, great, bye." Sam always made Ali feel a little better.

It was 1800, and they met up at the I-bar. Ali was already there when Sam walked in. "Hey, Sam," Ali said.

"Ali, what's going on? What did you want to talk to me about?"

"Admiral Callahan chewed me out today for arguing with Captain Smith about the medical board's decision. It sounds like he must have talked to the Admiral."

"Yeah, I'm sure he did if you argued with him."

"It seems like they've worked together all along! Probably that shrink Carver as well."

"What do you mean?"

"He interviewed me multiple times. Once at Sigonella and here twice! Some questions he asked me were unrelated to the crash incident."

"That doesn't surprise me, Ali; they do that in psychological evaluations."

"Yeah, but they sounded like how a cop would question a criminal suspect! The thing is, Admiral Callahan scolded and lectured me for it. Then, he nearly begged me to reconsider my resignation! He said he didn't want the Navy to lose me over it."

"He recognizes your skills and abilities and doesn't want to see you go, Ali, that's all!"

"It was probably more like he didn't want me to file an 1150 discrimination complaint against Smith. He would be implicated too if they investigated it!"

"Ooh, Ali, you don't want to do that! They'd never pursue a senior captain, or for that matter, a flag officer, just over your complaint."

"I know. I wouldn't bother doing it, anyway. It would never get resolved, and they'd railroad me out some other way for it. Gun-deck a bad Fit rep or something like that."

"So, what are you going to do?"

"Admiral Callahan suggested I put in for reserves. Just in case I changed my mind later. What do you think I should do, Sam? What would you do?"

"I'm with you on this one. I wouldn't be a happy Marine either if they grounded me permanently like this. I'm not sure how I would handle it either. Guess it depends on how devoted you feel."

Ali got a little upset and teary at him, her voice wavering, "I don't know how I feel about it right now, Sam. It's all I have! I'm still pretty upset at their decision and that fucking Smith!"

Sam sat a little longer with her, consoling her as best he could before he had to leave. Ali returned to her room at the women's BOQ and called Michelle. She tapped Michelle's number into her smartphone. Michelle answers sleepily: "Um, hello."

"Hey Michelle, it's Ali. How are you doing?"

"Oh, Ali! Hi, I'm so glad to hear from you! I'm mostly hanging in there. How are you doing?"

"Not so good at the moment."

"Wait, what? Why? What's wrong?"

"Listen, are you busy tomorrow? Do you have any plans?"

"No, why?"

"I'm taking the day off, and I'd like to come spend the day with you. I'll tell you all about it then."

"Okay, please do! I'm so looking forward to it already!"

"Okay, see you then."

The following morning, Ali drove to Michelle's place, out the front gate of North Island Air Station, across the Coronado bridge, I-5 North to Old Town—exit 19, right off the I-5 freeway, and a few miles further to Michelle's place off Mason and Jackson streets, near Presidio State Park. Ali loved that area, except for the highway noise from I-5 and the roads around Michelle's neighborhood. She rapped on the door a few times until Michelle answered.

"Hey Ali, come in!"

"Hi Shel, how're you getting along?"

"I'll live for now, I guess. Hugs please! "

"Okay, Shel," Ali wasn't so comfortable getting cozy with anyone, especially Michelle. Her attachment to her made Ali uncomfortable, but she dealt with it, considering their shared experiences.

"So what's the deal, Ali? You okay? You sounded a little rattled yesterday on the phone."

"Yeah, I'm pretty depressed right now. I got the ruling from the medical board yesterday."

"Oh, sounds like it didn't go as you wanted then. What happened?"

"They won't reinstate me to flight status. Then I got my ass chewed by Admiral Callahan for arguing with that Captain Smith, the medical board OIC!"

"Oh, that prick. I've had a run-in with him too! He pretty much said I was unfit for service any longer because of my headaches and other problems."

"Really? That SOB must hate women! He pushed us both out."

"In my case, Ali, he was right. I can't focus on much of anything. My head hurts all the time, and I can't remember shit a lot of the time." "Shel, you need to deal with that! What'd the VA do for you?"

"Not much. They sent me to therapy meetings with other combat vets and had me on Oxy for a while. Then they cut me off. The VA doctor said it's too addictive. So I had to find something else."

"What did you do, Shel?"

Michelle flashed Ali a smirking little smile and laughed, then said, "What do you think you smell, Ali? It's the sweet smell of colitas!"

"What?"

"You know, buds! I use CBD oil, too, but the genuine stuff works better."

"That stuff isn't good for you, Shelly!"

"Maybe not, but neither is a non-stop throbbing headache. Besides, I have a prescription for it. It's medical grade, so I use what works until I find something better or until the headaches stop."

"Don't overdo it, Shel. Did you try a civilian doctor?"

They told me the same thing. They'll prescribe this drug or that drug, but nothing solves it for me. What about you? What will you do now?"

"I'm not sure yet. I may request reserve duty just to keep my pension. I've only got four years before I can retire with a full pension."

"Why don't you just retire with a medical disability, Ali?"

"It's not that simple. On the one hand, they won't give me back flight status any longer, but on the other, I'm not, strictly speaking, unable to continue on active duty in some other less demanding job."

Michelle leaned back on her couch, thought for a minute, looked at Ali wistfully, and added, "I suppose it was better for me. They retired me outright, with a pretty sweet pension and a 100 percent VA disability rating."

"I don't know, Shelly; I haven't decided yet what I'll do. I'll have to think about it for a while."

They spent the day hanging out and reminiscing late into the afternoon about all their escapades and times together. They walked on the Embarcadero waterfront and lounged at the esplanade, one of Michelle's favorite places. They talked about everything that happened, looking out over the waterfront and watching the sailboats and the big carriers at the North Island piers across the harbor.

Ali didn't tell her it may be the last time she might get to spend time like this with her. She'd already made up her mind to leave the Navy. She felt too burned by everything that had happened. She put in her resignation request and continued working on the Admiral's staff until the end. That took three months to process.

During that time, she worked on duty assignments and stayed in touch with Sam and Michelle, but otherwise, she kept to herself on the Air Station when she was off duty, waiting for the time to come.

The official reason listed on Ali's 4510 resignation request was 'service-connected injuries and traumatic stress sustained in combat action.' In reality, to her at least, it was more about her feeling railroaded out of flying by a Senior Medical officer! One who she felt was opposed to women, or maybe just her in particular; she wasn't sure. She'd had enough.

Sometimes, the survivor's guilt was overwhelming. She knew those guys and felt responsible for them and her crew, Master Chief Brooks, the other Seals, and those young Marines who were lost while flying with her. "Why did I survive and have to deal with this shit?"

Then, it was time to process out, and it was all over. It was time to go home—back to San Marcos.

Ali dropped by for one last visit with Michelle before leaving San Diego. She thought about what she would say to Michelle on her way there. Ali knew Michelle would take it pretty hard, but it was time to go, and that was that. She walked up the stairs for what she thought would be the last time and rapped on the door several times, calling out her name.

"Shel, it's Ali. Are you home?" There was no answer; she knocked again, harder this time. "Shelly! Are you okay? Answer the door!"

Michelle answered finally, looking disheveled and hung over.

"Oh, hey Ali, come in."

"You okay, Shel, you look like shit!"

"Oh, sorry, I'm tired, and my head hurts, but I'm okay, I guess."

"You look messed up. Are you overdoing those drugs, Shel? It looks like it's fucking you up a lot!"

"I know, I can't shake the headaches! I don't know what else to do!"

"Try finding a new doctor, for God's sake! Shel, I stopped by to let you know I'm out of the Navy as of today. I'm planning on heading home for a while. To Texas."

"What? Please don't leave me, Ali, oh God!"

"Come on, you knew I wouldn't be here forever. We'll talk when you want to. You can call me anytime you want to or need to."

They spent most of the day together, Michelle whining about being left alone and Ali trying not to feel guilty.

"Shelly, I'll be there for you if you need my help. I have to go now, okay?" Then she left, and that was it. Michelle never saw her again.

That evening, Ali met Sam at the I-Bar for a farewell drink and a private chat. Sam walked in and noticed her sitting in her usual place at the bar.

"Hey, short-timer! How's it feel?"

She answered him, sounding tired and depressed, "Oh shit, don't ask me that, Sam. I can't even believe it myself! After all of this crap, this is how it ends. I can't even process it. I stopped in and spent some time with Michelle today. She's practically hysterical over me leaving."

Sam looked at her with a shocked expression and asked, "Leaving? Where are you going?"

"I'm going home to San Marcos, at least for now, and try to figure out what's next for me. I had a place there that my dad left me. It was his mother's house he grew up in. I have to decide what I'll do with that too."

"Will I ever get to see you again? Now I'm depressed too, Ali! I'm crazy about you, you know that! Damn, this sucks, girl, you know!"

"I know you do, Sammy, but I've told you I'm just not up for that. Especially now. I don't know where I'm going or what I'll do next."

She tried to sound encouraging, smiled at him, and said, "You're a good man, Sam. We'd be good together. You'll be the first to know if I change my mind." Then she added, "Even though you are a big-mouthed Jarhead!" They both got a laugh out of that one.

Chapter 10 - San Marcos

Another Life

Ali, now resigned from active duty after the medical board fiasco, hadn't yet decided about reserve duty. Uncertain about what would come next, she was back home in San Marcos. It was time to wrangle with the VA over disability benefits.

Since she'd lost part of her leg, she applied for and received a 100% disability rating, ultimately getting it. They wanted to rate her at only 75%. Fortunately, she'd found an excellent benefits attorney who argued her case and eventually got her a 100% disability rating.

The sheer amount of paperwork and the bureaucracy at the VA were overwhelming. The process seemed daunting and endless, with many forms and many appointments. She soon discovered that she was not alone in her experiences.

They provided continued and good-quality medical treatments, but less of what Ali felt was enough psychological support. There were group meetings with other vets like her. Ali served with unwavering dedication during her Navy career. It just ended in a retirement check, endless therapies, and failed dreams.

She wasn't happy about it—a real rollercoaster ride of emotions, triumphs, and challenges. Life took a few unexpected turns after that. The biggest challenge was the emotional toll. Adjusting to civilian life with a prosthetic limb proved no easy feat. It never is, even when you are all there in one piece.

Despite all that, Ali encountered other roadblocks along the way: delays in necessary medical services, problems with scheduling appointments, and setbacks that tested her patience and caused more emotional distress.

She knew she had to keep moving forward or never feel normal again. She just had to keep thinking they were doing their best to help her and countless other veterans like her. Throughout those times, Ali found solace in connecting with other veterans who'd gone through similar experiences.

The VA sponsored support groups and counseling sessions, providing a safe space for veterans to share stories, seek guidance, and find camaraderie. Experiences that proved invaluable and made her realize she wasn't struggling alone. It proved to be a strong support network. As with many things, sometimes it just wasn't enough.

It was a mix of highs and lows. The VA's challenges and frustrations happened regularly. Many very dedicated and committed professionals, many of whom are veterans themselves, do their best to take care of fellow veterans at the VA.

An Unexpected Visitor

On her way home after one of those group therapy sessions, she reflected on her experiences since the crash, some with gratitude for those who did their best to help, some with trepidation for all the problems, and others for people like Smith, who put roadblocks in front of her. In the middle of her reflection, her phone buzzed and rang like an alarm signal for a catastrophic highway accident: "Yes, hello."

"Ali, it's your Uncle Collin."

"Uncle Collin! Hi, oh my god, I haven't heard from you in ages. How are you doing?"

"I'm fine, sweetheart. How are you, though? I heard about the accident and that you'd finally made it home. I'm so sorry I didn't contact you before now, but I didn't know where you were or how to get hold of you. I would've been there for you if I had known where you were. I've been worried for you since I heard about the crash."

"Well, thanks for that. I'm okay now, more or less, I guess. I'm just trying to adjust to everything now. I have good days and bad days, you know."

"Ali, I'll be in San Antonio next week. I want to stop by and see you, if you don't mind."

"Sure, of course! I'd love to see you again! But what are you doing in San Antonio? Don't you work in D.C. now?"

"I do, but I'll be in San Antonio for a business meeting with the Air Force. Thought I'd make a side trip and come see you."

"Sure! I'd love to see you! But how did you get my number? I haven't seen you in ages. I don't think I ever gave it to you."

"Well, Ali, I still have many connections with the Navy. After I heard about your accident, I started calling around, trying to find you."

Ali's uncle Collin, a retired Navy Captain, Collin Yves Martel, her mother's brother, worked for a defense contractor out of their Washington D.C. offices. He'd been a combat pilot during the Gulf War and was now a contract administrator and sales liaison to the Air Force.

He'd read a lot of chatter about the goings on in Syria when her helicopter crashed. The reports made him worry about her being involved, so he dug a little deeper and found out it was her who had crashed and had been severely injured.

A few days after his initial phone call with her, he'd finished his business in San Antonio and started heading for San Marcos to see her and spend some time with her there. He tapped her number into his phone as he started for her place. Ali's phone rang and buzzed away like it would catch fire if she didn't answer it immediately.

"Yes, Hello."

"Ali, It's Uncle Collin. I'm taking off from San Antonio now. Are you still up for that visit, or is this a bad time?"

"Hi, Uncle Collin. Sure, it's fine. I'm just here by myself, limping around Daddy's house."

"Ali, it's your house now. Your dad left it to you. Don't pile on yourself."

He could hear her choked up a little, voice wavering.

"I know. I miss him so much, you know!"

"I know. We all miss him, sweetheart. I'll be there in a couple of hours. We'll talk about it then. See you soon."

"Okay, Uncle, see you then, bye!"

Collin influenced her desire to fly. He started flying F-14 Tomcats in the 1980s, later moving to F/A-18 Hornets. As a young girl, she loved watching them fly with him and her dad at airshows. Then, everything came undone when her dad passed away—it all stopped.

Collin's driving time was quick, driving to her place. The trip up I-35 was easy, and traffic was light that day. He rolled up in front of her house, parked his rental car, walked up to the front door, and knocked on it a few times until she answered. "Hi! Uncle Collin! Come on in! I'm sorry it took me so long to answer the door. I don't get around as fast as I used to on this stupid prosthetic leg."

"Ali, I'm so sorry I couldn't get here sooner. When I heard you were back home, I started trying to get away to see you. I'm sorry it took solong. How are you holding up, honey?" Collin was a little emotional at this point. He'd been close to Ali and her dad when she was younger and felt he should do more for her now.

She could sense him trying to keep his composure, hugged him, then said: "Don't feel like that; it's not up to you to look after me now! But thanks for the thought. I appreciate you stopping by. I don't get many visitors these days. It's like I have leprosy or something! This damned ugly prosthetic leg makes people not want to even be around me or talk to me! I don't know what I'm going to do!"

Their conversation had gotten emotional and triggering for her now. Ali just started coming unglued and crumpled on the couch beside him. Then it all rolled out like a river flowing over a waterfall.

"I don't know what to do now! Who'll ever want me like this?"

Collin sat there next to her, cradling her like he'd done when her father passed years before, trying to comfort her through it. "It's okay, sweetheart. Things will smooth out. You'll figure it out. You'll see. You're still a strong, beautiful woman with many years ahead. It'll just take time, that's all, just a little time."

Inside, though, he hurt as much as her. He had no children of his own, so she'd always been the apple of his eye. It killed him to think of her, his niece, such a bright, beautiful woman being disfigured like this. Ali was indeed a lovely woman. No one ever understood why she'd joined the Navy.

Collin knew she'd done it because of him. As attractive as she was, she could have done anything else and been successful at it, but she chose to be a naval officer and a pilot like him. She'd paid a hefty price for it. It killed him inside. He felt responsible somehow for the whole thing.

"Ali, I was so proud of you for pulling yourself together like you did after your dad passed. Then that nonsense with CPS, the foster homes, and you taking off like you did. It seemed like you were so lost. I felt so bad about it. I wanted to help, but couldn't figure out how to make it work for you or me. I was always off on deployments somewhere. Then, you just picked yourself up and overcame everything!"

"I knew it was because of me you joined the Navy. I was very proud of you for it, but terrified, too. I was so afraid something like this would happen. I felt terrible about it and responsible."

"It's not your fault either, uncle. If anyone is to blame, it's me. I should have expected something like that would happen, considering where it was."

Collin reflected a little and told her a story. "One of the guys I flew Tomcats with during the Gulf War was shot down over Iraq. Nobody was ever sure exactly what happened to him, but all the evidence pointed to him being hit by a missile from an Iraqi Mig."

"He was listed as MIA at first. In the end, since they never found out what had happened to him, his status changed to KIA/BNR (killed in action but never recovered.)

"Those of us who knew him kept looking for him and pushing for the search to continue. They even sent in a Seal team to search for him. They never found him until a bunch of Marines found what was left during OIF years later."

"I had nightmares that something like this might happen to you! Now, here we are, and it's nearly happened! I'm just so happy and thankful you survived, Ali! You'll get through this. I know you will!"

"But where do I go from here, Uncle Collin? I can't fly any longer. I don't even know what else to do!"

"Well, you know I'll help however I can, but you'll have to decide what's next. Nobody can do that for you."

Thoughts of Suicide

Ali had been through a lot up to now: a year in two different military hospitals, recovering from a broken back and losing her leg, physical therapy, and then the Balboa prosthetic clinic in San Diego, learning to walk again on it. She hated it; "that damned peg leg!"

She knew she'd have to accept her new reality, though. Back home in San Marcos, she spent much of her time alone now, going through more therapy with the VA, working out, and trying to keep up as much of her strength and mobility as possible.

She spent long periods by herself dealing with it. Often thinking about the others who hadn't made it. Like Master Chief Brooks. She and Michelle had both seen him splattered all over the desert near them when trapped in their wrecked helicopter. The guilt over it ate her up inside. Those were the worst days - as if she could have kept it from happening.

She blamed herself for not being smarter and considering all the possibilities when planning that mission. Ali never accepted for a minute that there was no way she could have known or avoided the ISIS assault.

"I should have been smarter; why wasn't I smarter? Why did I make it when the guys with wives, kids, and families didn't? What about Michelle? The poor girl was ruined by it. I hope she's okay and hasn't overdone all those damned drugs."

It haunted her constantly. The survivor's guilt painted her thoughts and her heart, often bringing her to tears thinking about it, trying to cope. The worst part for her now was no longer being able to fly, the one thing she loved doing above all else and worked so hard to do.

She'd spent years putting herself through college, scrimping and scrounging to make a living and pay for school. No family money or help from anyone was ever there. After her father's death, her mother lost her ability or desire to deal with Ali.

Her Grandma stepped up after that. Abuela, her dad's mother, was a very strong and caring woman who raised Ali well. She gave Ali the solid ethical beliefs and family and cultural ties she needed then. Ali's life with her was a happy time. Those years with her proved to be the source of her inner strength and gave her inspiration, grounding, and a drive to excel.

The happy times with her went by too fast as Abuela died suddenly a few years later, leaving her alone again and on her own now. Since she was still only 15, the family courts placed her into foster care. Ali had many problems then, moving from one foster home to another.

Finally, Ali had just had enough and ran away. A pretty young girl on her own, she quickly learned its 'realities.' Many a night, she spent alone, cold, hungry, huddled in some abandoned house with a few other runaway kids like herself, or camped out like a railroad hobo. Even that didn't break her.

As strong-willed, savvy, street-smart, and determined as she was, Ali was still vulnerable. A few good people along the way helped when they could. Some gave her a safe place to sleep and fed her. Others, of course, were not so honorable. She learned quickly how to spot and avoid these types.

Now, in this time of extreme personal loss, trauma, and pain, all the old feelings and resentments from those times had come back to haunt her. Everything just piled on and became unbearable. She felt overwhelmed and wanted to die. Now living in her father's house in San Marcos, she'd set aside her old childhood bedroom as an office and study.

It was decorated with displays of all her college and Navy memorabilia: pictures of friends, some who were already gone, some flight gear she'd 'appropriated,' a shadow box full of all her military patches, a four-row ribbon stack, awards, and keepsakes, the Purple Heart that had been left pinned to a pillow on one of her hospital beds while she was out of it, and an American flag.

That day, she'd hit rock bottom, in a deep depression, feeling like it just had to end. She felt no one cared, like when she was young and alone, with a few exceptions:

"At least I have Uncle Collin, Michelle, and Sam."

Ali had had it. She wanted no more, no more loneliness or feelings of guilt and loss. She stood up, went to her bedroom closet, got out her Dress-White uniform, and put it on—full dress, ribbons, and medals. Ali never got out of the habit of keeping them up to inspection standards, as if she would ever need them again. Everything was clean, pressed, starched, and spit-polish perfect.

She put it all on and eyeballed herself in a full-length mirror and did a little of a 'personnel inspection' on herself. A sad, ironic feeling enveloped her. She couldn't help but think how awkward-looking it all was with that prosthetic leg.

It just made things worse, seeing herself this way now. Teary-eyed, depressed, and feeling like she was at rock bottom. She walked over to her nightstand, opened the top drawer, unlocked a small gun safe, took out a government-issue Colt 1911 service pistol, loaded it with a full mag, and headed back to her office study.

She took the flag down from its mounting, folded it in half, sat in a brown leather-covered recliner, and draped the stars and stripes over her left shoulder like a sacred security blanket.

Sitting there with the pistol on her lap, wearing those perfect dress whites, a pair of clean white gloves, and the stars and stripes draped over her shoulder, former Lieutenant-Commander Althea 'Ali' Cabrillo caressed the stars and stripes a little with her right hand, held the pistol in her left hand, and was going to do it! She would 'end it' then and there!

She sat for a long time thinking about it, weeping and sniffling a little. She didn't have any of her own family to miss except Uncle Collin. Nobody else she thought would miss her except for Sam and Michelle.

Military life can do that to a person. You're always gone and never have much time for family things, dating, keeping up relationships, or even your own children. Thoughts of Sam, Michelle, and Collin crept into her thoughts, reasons not to do it.

"Oh god! Michelle would be hysterical. She'd go catatonic if I did this! Do I want to give up like this and take a coward's way out? I don't know, shit, I don't know! What about Uncle Collin? It would break his heart and poor Sam, the best man I ever knew!"

She sat there sobbing, wrestling with herself and brooding over it for a while, then her phone buzzed and rang. She looked at it, not recognizing the number, but answered it anyway. The phone call came at the right time. It broke her desperate mood 'for the moment.'

She got hold of herself enough to answer calmly: "Hello."

"Yes, hello. Is this Lieutenant-Commander Althea Cabrillo?"

"Well, I was her, yes, now I'm just Ali Cabrillo. Who am I speaking with, please?"

A New Friend

"This is Colonel Jean Schaefer, retired U.S. Air Force. Your VA friends thought you might like to chat with someone."

"I understand you're having difficulties adjusting to civilian life with your prosthetic leg. They've not seen or heard from you for a while. So, they asked me to call you and see how you're doing. Are you all right? You sound upset."

Sniffling a little but keeping her composure, she answered, "I'm fine, thanks for asking. Okay, but why you? Why would they ask an Air Force Colonel, retired or otherwise, to contact me?"

"Well, that's complicated to explain over the phone. Can we meet somewhere for a coffee? It'll be clear enough then."

"I don't know, Jean. I'm unable to get out right now."

"Come on, Althea! It'll be good for you to get out for a while and chat with me. I can tell you're in a bad way right now. We can meet at a coffee shop called the Daily Grind. I understand it isn't too far from you, and the coffee is supposed to be excellent there. Say an hour?"

Ali thought for a minute, "What the hell? I was just going to shoot myself, anyway. I guess that can wait for a while." So she accepted Jean's invitation.

"All right then, I suppose I can do that. Some fresh coffee sounds good."

"Great, see you then."

"But I don't know you. How will I recognize you, or you me?"

"Don't worry about that. You'll recognize me and understand why I want to chat with you."

"All right then, I'll see you soon. Goodbye." Ali thought about it for a minute; "Sheesh, that was damn strange, a full bird-colonel. What the hell would she want, anyway? Why not? Now I'm curious."

That call from Jean was just what Ali needed to diffuse her depression and pull her out of a serious funk. Just enough not to pull the trigger on that Colt 45. Her meeting with Colonel Schaefer proved to be a significant turning point.

Ali put aside the flag and pistol, returned to her bedroom, changed out of her dress whites into street clothes, and headed off to meet Schaefer. A little while later, she pulled up in front of the coffee shop, parked her car, and walked into the coffee shop. Schaefer was sitting near the front door.

The coffee shop

Ali noticed her right away. She was pretty hard to miss. Schaefer had lost one of her legs in Iraq. "Hello, Althea; I see you've noticed what we have in common."

"Yes, I see that! Nice to meet you, Jean; what happened to you?"

"OEF, RPG, pot shot from outside the perimeter wall. I was intel, working out of the green zone in Baghdad in 2005. Ali Babba shot lots of them at us back then. One hit near enough to cause this, and the rest, as they say, is history. I understand you were a Navy pilot, weren't you?"

"I was, but officially, we're known as Naval Aviators. ISIS shot my helicopter to shit in Syria a couple of years ago now, and well, I spent a year flying a damned hospital bed with a broken back, lost my leg, and another six months in physical therapy, learning to walk again on this stupid prosthesis!" Even as she said it, the thought upset her all over again.

"I understand how you feel, Althea. My wounds ended my Air Force career, too. I went through the same kinds of things you are dealing with now. That's why they asked me to meet with you and see if I could offer you help and encouragement. You learn to live with it, you know. You need to realize you were lucky to have survived!"

Ali knew Jean was trying to help, but Jean's words didn't make her feel very lucky. She was just getting worked up, emotional, and teary-eyed, flailing her hands up and down, looking like she wanted to take off and fly like a bird.

"Oh shit, Jean! I don't know what I'm doing now. I was doing everything I wanted to do my entire life, doing everything right! I worked so hard for it! Now it's all gone, and it's never coming back!"

"Well, my solution was to find an alternative. I went back to school and got my PhD in psychology. I wanted to help other vets like you cope with it. It turned out it was helping me cope as much as helping others deal with their traumas. Think about something like that. You could be a big help. There are never enough therapists for this."

Ali regained her composure a little, took a breath, and replied.

"I don't know, Jean; I'm not sure I'm cut out to be a counselor. I don't know, maybe. I'll give it some thought, though."

"Well, it doesn't have to be a therapist, although there's a great need for more. I'm trying to point out that you must decide to move on with your life and do something else. You have options, you know. Some other government job, maybe.

That might be the way to go. You've got a serious set of skills."

"You might use them as a flight instructor or in a civilian contracting job. You might even want to try something creative—art, writing, something like that."

"Well, maybe. That sounds interesting. It's something to investigate, anyway. Thanks for asking me to meet with you here today, Jean."

"Sure, I hope this helped. We could do this again anytime if you want to get together and chat. I've also got someone I want you to meet, anyway."

"Really? If it's a man, I'm not up for that right now, if that's what you are suggesting."

"Oh no, he's not a man, but he's a disabled vet, too."

"What? You just said it's not a man. What are you even talking about, Jean?"

"Tell you what; meet me here tomorrow, and I'll introduce you." Jean's curiosity-generating subterfuge piqued Ali's interest further. Shaefer clearly wanted her to agree to meet again.

"Okay, then, you're on! You've got me curious."

After this chat, Ali felt much better and began throwing a curious grin of anticipation. She never mentioned to Jean that her phone call had maybe saved her life that day, but she still hated that "fucking peg-leg!" just not so much now!

Ranger

The next day, Ali returned to the coffee shop and met Jean again. This time, Jean had a friend with her—a hairy, panting friend!

"Hey, Jean! Who's this? Well, hello there, puppy! Who are you?"

"Hello again, Althea. This is Ranger. He was a military service dog until he lost part of his left front leg and paw in Syria. I'm unsure if he was attached to a unit in the same area as you, but he was there."

"Well, Ranger, I guess we have something in common too, don't we, boy."

Ali patted Ranger on the head, stroked the fur on his back a few times, and scratched him behind his ears. Ranger looked up at her with those wide, inquiring eyes that Shepherds use when trying to decide if you're a good guy or a bad guy they want to bite.

He wagged his tail a little at the attention he was getting. Ali loved dogs and took a shine to Ranger right away. Jean picked up on that quick time. Ranger seemed happy to meet Ali as well. Jean thought to herself, 'Duly noted!'

It seemed like a pretty good segue into the next part of Jean's conversation. "Would you ever consider adopting a disabled war dog like Ranger Althea? He's available and needs an understanding friend, too."

"Really? Where did you find him? What do you mean, understanding?"

"There's a retired service dog adoption center at the Air Force base in San Antonio. I've been fostering him while they try to find him a permanent home. I said 'understanding' because Ranger also suffers from traumatic stress from combat. Dogs get it, too. He's kind of withdrawn and nervous around people he's not familiar with."

"He also can't deal with loud noises anymore since his wound occurred. He thinks they're explosions, I guess. You'd have to give him time to adjust to your home and feel safe with you. He'll be a little withdrawn until he accepts that you really want him. You would also have to apply at the shelter to adopt him. They have an orientation process they put prospective owners through."

Ali was already hooked and loved Ranger right away. She didn't want to seem like she was that easy to convince, so she made a little of a fuss to make it look good.

"Jeez, I don't know, Jean. I don't know if I can commit to that right now. I have a hard enough time taking care of myself. I love him, though! He's such a pretty pup, aren't you, boy?"

Ranger wagged his tail a little, thumping the wood floor of the cafe with it. He loved her attention.

"He seems to like you. He rarely warms up to anyone right away. If you aren't sure, Althea, you could try fostering him for a while to see if it works out for the both of you."

Ali smiled at the thought. "I'll give it some thought, okay?"

Ranger

"Great. Don't think too long, though. He might get adopted by someone else!"

"I'll let you know in a few days, okay, Jean? I want to give it some thought. Goodbye, Ranger, you pretty boy!"

"Okay, that's fine. Just call me. You've got my number."

Jean already knew she'd caught a fish. Ali loved Ranger and needed a companion. It's not good to be alone all the time, especially when you're dealing with traumatic stress, whether for dogs or humans.

It didn't take her long to decide on Ranger, maybe even adopt him permanently. She hadn't had a dog since she was a young girl. The thought of taking care of a handicapped veteran dog who needed someone who understood its problems too appealed to her, so she made the call.

"Hello, this is Jean Schaeffer."

"Hey Jean, it's Ali Cabrillo."

"Hi Althea, I didn't expect to hear from you so soon. Have you decided about Ranger yet?"

"Yes, I've decided I'll take Ranger and see how it goes."

"Great! Would you like to take him today? I'll take care of it with the adoption center if you like. If you decide to adopt him permanently, contact them when you're ready."

"Can I? Oh, that would be great, Jean. I'm looking forward to spending time with him now!"

"Outstanding, Althea; I'll pack his gear and bring him right over if you're ready for him."

"That would be great, Jean. Bring him over as soon as you're ready."

"Great! Please give me a couple of hours to get everything together, pack out his gear, and drive up there. See you soon."

Jean ended the call, feeling satisfied her plan had worked perfectly. She smiled and thought to herself, "Mission accomplished!"

Later, Jean drove up to Ali's place. Ali spotted Jean's car through her front window and hurriedly opened her front door, waving at Jean and Ranger and welcoming them both. Jean opened her SUV's tailgate to let Ranger out. Ranger couldn't hop out of it like most dogs would, so she picked him up and helped him out.

After she put him down, Ranger headed for Ali, walking toward her with a limp but tail wagging all the way. Ranger sat beside her, leaning against her, whining like a little pup, tail wagging and thumping on the stoop in front of her house, excited to see his new bud again!

"Jeez, Ali, he's taken a real shine to you right off the bat. I've never seen him warm up to anyone like that so quickly!"

"Yeah, I guess he recognizes a kindred spirit. Don't you, boy!" Ali smiled and scratched Ranger behind the ears.

"Heck, you two will get along famously. Look at him; it's like he's found a long-lost friend!"

"Yeah, I'm amazed myself! How are you, Ranger?"

"Okay, Althea, let's get his stuff out of my SUV and take everything into your place."

Jean and Ali unloaded everything from her vehicle and carried it into Ali's front living room. Once they'd finished, Ali and Jean sat down and chatted for a while. Ranger sat next to Ali, pressing up against her. As she rummaged through the dog's gear, she stumbled on a medal presentation case holding a purple heart, prompting her to ask Jean about it.

"What's up with this, Jean?" Ali had to know.

"It's his Purple Heart medal."

"What? How is that even possible? The military doesn't give awards to animals, even if they deserve them!"

"True, but his last handler and the people at the adoption center put it together to commemorate his retirement and service. It's not official, of course, but they thought he deserved one. There's also a certificate, just like the one they gave us for our medals."

"I didn't know you were a war hero, Ranger! We'll have to put this on the wall beside mine!"

Ali smiled a big, pretty grin at the dog like she was a proud new mom. He looked at her with those big questioning Shepherd eyes they look at you with that make you love them even more, then thumped his tail on the floor for good measure. Jean looked at her and just grinned a little to herself.

"Ali, you've already decided to keep him, haven't you?"

"Yeah, I guess you figured I would anyway, didn't you?"

"I didn't know for sure. But I didn't doubt it either."

Ali grinned. "Am I that easy to read, Jean?"

"Yes, Althea, I'm afraid you are!" They both got a laugh over that.

Ali was sympathetic towards Ranger. She had a weakness for dogs anyway, so it didn't take but a few days for them both to feel entirely at home. The two of them quickly bonded and became inseparable.

She realized how much she needed him, as much as the dog needed her. They became an inseparable pair and always stayed together. When he wore his veteran war dog vest, Ranger could accompany her everywhere she went. People loved seeing him wearing it.

Ranger was Ali's personal guard dog. He became very protective of her. With that old war dog by her side, she never had to be concerned with anyone trying to mug or molest her. Ali wasn't the sort to worry about that, anyway. She was a trained ex-Navy pilot, skilled in martial arts and the self-defense training she learned at SERE school, and in her early youth.

Even with that prosthetic leg, she was still a formidable woman who knew how to protect herself and still packed that 1911 Colt 45, too—the one she'd nearly shot herself with. She knew how to use it very well. Ranger just added a bit more to her confidence. Ali took care of him for the rest of his life. And that's how Ali met Ranger!

Chapter 11: Collateral Damage

Michelle's Story

After Ali, Michelle, and the others were medevac'd to the ship, and Ali was sent to Sigonella for emergency surgery, Michelle spent the next few weeks recuperating onboard. The doctor, Commander Slatter, ordered her to rest and see if her headaches improved.

His biggest concern was her concussion symptoms and possible T.B.I. (Traumatic Brain Injury). She showed many signs of having one or both. The signs and symptoms can be subtle and may not show up immediately. They can last for days, weeks, or even permanently.

Michelle never overcame the headaches and problems with her memory and coordination, as Slatter had hoped she might. After a few weeks, at her next medical check-up, Slatter let her know she would be departing.

"Lieutenant Robbins, how are you coming along? Are you still having problems with persistent headaches?"

"Yes, Doctor, I still have them, sometimes worse than usual."

"Has your memory improved at all since the last time we chatted?"

"I'm afraid not, Doctor. I can't concentrate on anything, either."

"Well, I'm sorry to hear that. I must inform you that you'll be detached from your squadron and sent to Landstuhl, Germany, for continued treatment. I've gone as far as I can here."

"Why not Sigonella? Isn't that where Lieutenant-Commander Cabrillo was medevac'd out to?"

"Yes, but doctors there indicated they didn't have facilities to treat your condition. So, the decision was made to send you on to

Landstuhl instead. You'll be medevac'd to Sigonella and transferred on to Landstuhl."

"Oh, when will that happen?"

"I don't have that set up yet. You may assume it will be the next COD or Medevac flight. If not, then shortly after that. Your squadron will forward your gear to you. You don't need to take anything but essential clothing, uniforms, Military ID, and records. If you have a passport, you can take that as well."

"Yes, sir." Michelle didn't care if she wasn't going where Althea was; it didn't matter, but traveling with her head throbbing as it made her uneasy. They were nearly continuous unless she was doped up.

She wasn't too concerned about leaving the ship, either. Since Ali was gone, she didn't want to be there, anyway. Her trip through Sigonella was short and uneventful. She didn't care; Michelle just wanted to get rid of the headaches.

An Air Force Medevac transport picked her up there the same day, heading for Landstuhl, Germany. Her time there was short as well—just a quick stopover on a trip home…

"Lieutenant Robbins, we need to talk," the attending doctor began softly, pulling up a chair beside the bed. "Because of some complications, we're transferring you to Balboa Hospital in San Diego."

Michelle's breath hitched. The Doctor's statement confused her.

"Is it serious?" she asked suspiciously, her voice barely above a whisper.

The doctor met her questioning gaze; "We believe they have more specialized care for you there that will better address your needs. It's precautionary, but you'll be in good hands. It's your home of record as well."

Reality settled in. She felt a wave of apprehension—primarily suspicion mixed with anxiety but hope that things would improve. The thought of a new place, new doctors, and, best of all, being near home was comforting. "When do I leave?" she asked, trying to keep her voice steady.

"Soon. You'll be on the next flight back to Conus, probably within the next few hours. We'll make sure you're comfortable in the meantime."

After that conversation, she felt better about it, now looking forward to the trip. Hoping Balboa could help her.

Medical Discharge

Naval Medical Center San Diego (NMCSD), 'Balboa Hospital' to most sailors and Navy veterans, or "The Pink Palace": Captain Smith, head of the Neurology department, discussed Michelle's condition with her.

"Lieutenant Robbins, as you know, we've been doing a lot of tests on you these last few weeks."

"Yes, Captain, I'm aware of that, sir, so what are we talking about then?" she asked.

"We've not been able to determine with any certainty if your continued severe headaches are because of concussive trauma from the crash, T.B.I., or something else. There's no medical evidence of physical injuries over and above the concussion symptoms."

A sinking feeling overtook her now. She felt what was coming next. "So, what are you telling me, Captain?" she asked tersely.

"Lieutenant Robbins, the medical review board has reviewed your case thoroughly. Since you continue to have these problems, the ruling has been made to release you from active duty for medical reasons."

Michelle was dumb-struck! "So, so that's it then, just like that?"

"I'm afraid so, Lieutenant. In your condition, you won't be able to fly any longer. The review board has determined that your continuing headaches and other physical concerns would preclude you from effectively accomplishing any duty assignments for the Navy."

He continued, "After your separation paperwork is completed, you'll be referred to the local VA for additional follow-up treatments, but it will be up to you to contact them and make your initial appointments."

Stunned but not surprised, Michelle's response wasn't emotional, but she was crestfallen; "Yes, Sir, understood, Captain."

"Good luck then, Lieutenant, and I'm very sorry."

She doesn't say it but thinks, "Yea, fuck you too, Captain asshole!" After all this, she wanted nothing to do with the Navy! In her mind, she was more worried about her bestie, Ali, than she ever was about her Navy career.

Michelle and the VA

Michelle's next stop was the VA Medical Center in Mission Valley. The VA does its best with the resources it's given, but Michelle found them to be of little help. There was never a doctor available during her visits. They always assigned her to a nurse practitioner, who couldn't do much other than prescribe painkillers for the headaches and the pain in her aching arm.

What Michelle resented the most was they always asked if she ever had thoughts of harming herself or others. As far as she was concerned, her arm and her head hurt all the time with those painful and debilitating headaches. There weren't any solutions other than drugs and group therapy sessions, which she hated. After a while, she just stopped going. It was clear it was going to be painkillers from now on!

Desires and Addiction

Michelle had been on a roller coaster ride for some time now: the crash, the medical treatments, the psych evaluations, and the VA. She wasn't improving and had shut out most people and activities. Her symptoms continued: headaches and pain in her left arm, **problems** with monthly cycles and her memory, and trouble with her balance. She felt disconnected from what most people would consider an everyday life.

In her desperation, feeling like she just wanted to escape the lingering headaches, pain from her other injuries, and desperate feelings of loss and isolation, she looked for something more potent than the meds she had been using.

The VA had already cut off her painkillers. They wouldn't renew the script any longer. Telling her the drugs they'd been giving her were "too powerful, addictive opiates." For her, it was the only thing that helped all the long months since the crash. Now she was going through severe withdrawal symptoms. "Please give me something! I'm still dealing with pain and headaches!"

She isolated herself more and more from social contacts. That didn't help her mental state, either. She was panicky for anything to ease the pain and not think about her anguish over lost friends, most of all Ali! "Where is she? What happened? Is she okay, sick, alive, or what?" Desperation drove her to hit the streets and go looking for alternatives, an under-the-table source of OxyContin.

Her mental state didn't let her realize a lot of her pain was the pain of opiate addiction and withdrawal. The pain in her arm and the headaches, "The goddamned headaches!" just made things worse.

The Navy doctors were convinced she'd suffered a concussion in the crash, and that caused her continuing headaches, but they couldn't come up with substantial evidence for it, or a treatment that might resolve the problem for her.

Carlos, the therapist

Walking along the embarcadero looking for a connection, one of her favorite places in San Diego, trying to come up with what to do, she sat down at the esplanade. She'd been sitting there for a while when a stranger approached her. He approached her and said: "Hello, are you Michelle Robbins?"

"What, who are you, and how do you know my name?" she asked suspiciously.

"My name is Carlos. I work with your Veteran's support group. They're concerned you haven't been coming around for a while. Nobody has seen or heard from you, and they want to make sure you're okay. They gave me your contact info and asked me to contact you. I went to your home earlier today to see if we could chat a bit if you had some time."

"Since you didn't answer, I assumed you might be here. The support group members told me you enjoyed the Embarcadero waterfront and watching the ships come and go from North Island and the harbor. They described you to me, so I took the chance it was you sitting here."

Michelle didn't know what to make of this dude other than to respond with, "So what do you want - leave me the fuck alone, man!"

He was surprised at her withdrawn, angry response, but replied in a friendly, low-key manner.

"Okay then, sorry to bother you. If you change your mind, I'll be around. You can find me through your veteran's group. Here, I'll leave you one of my cards. Call me if you decide you want to talk about anything."

"My number is on the card. I can help you deal with this, okay? But you'll have to decide if that's what you want to do. Are you sure you wouldn't want to chat a bit?"

"Not unless you can hook me up with some Oxy. Otherwise, no. Thanks for the offer, though."

Michelle took his card but wasn't paying much attention to him. She replied, "No, I'm not in the mood for a chat, man. Not unless you can hook me up with some Oxy. Otherwise, no, but thank the guys for their concern. I'll let you know if I ever want to talk about anything!"

"Okay, fine. Take care of yourself. Your friends at the support group are very concerned for you. Let them know you're okay or make it to a meeting now and then."

Her response was unconvincing, but clear enough: "I'll think about it. You can tell them I'm still alive!" Carlos didn't have a response for that. Carlos realized she didn't want to talk or open up to him, so he just left her sitting there. Walking away, he made mental notes to himself, "displaying classic symptoms: irritability, anger, apathy, depression".

He knew Michelle was having debilitating headaches. He wondered at the rest. "Did she have flashbacks or nightmares? The crash, the battle, or both? Maybe even other experiences."

He shrugged and sighed a bit at her reaction. Carlos was a licensed therapist who worked voluntarily with veterans and their support groups through the VA. He was also a veteran of Afghanistan and knew the signs well when he saw them.

After leaving the Army, he pursued a career in psychology and became a licensed therapist. He wanted to understand better his feelings and the effects of combat stress on others. He was driven to provide support to other veterans in every way possible. Her response to him left him feeling frustrated, as she appeared to be lost and in pain.

Carrie Lambert

Sometime later, Michelle meets a new 'friend,' Carrie Lambert. She was sitting on a bench along the Embarcadero esplanade the same afternoon that Carlos found her. Carrie could be described as an opportunist, a sociopath, or a grifter who recognized a mark when she saw one. She zoomed in on Michelle's apparent need for 'something,' a drink, a fix, companionship, or 'something else.'

Michelle would never admit it to herself or anyone else, for that matter. Inside, she was pining away for Ali, the one person she loved the most in the world. She feared she'd never see or hear from her again after Ali retired from the Navy and left for Texas.

Ali was gone, living in San Marcos, Texas. Michelle had little contact with her now. She tried to stay in touch, but Ali was dealing with her problems, too. They just drifted apart as time passed.

Carrie was a tall, willowy blonde woman in her late twenties. She looked, for all the world, like most people would think of a California beach girl: pretty, with long blonde hair, big blue eyes, and a golden-brown tanned body—a natural beauty.

She knew people were drawn to her for her beauty and amiable nature, so she worked it to her advantage. Michelle was utterly captivated. "Wow, she's so beautiful and friendly," she thought, letting her desires run wild. It never crossed her mind to question whether or not she was being hustled.

Michelle was no slouch in the appearance department either; she was still in good shape from her service life. Naval Aviators are well-conditioned physically. Even in her current diminished state, she kept her appearance up. It didn't hurt that she was pretty, either.

Michelle had medium-length, light tawny chestnut brown hair, hazel-colored eyes, and a wisp of freckles on her face. She was not so much model material, but attractive in her own right. Carrie, on the other hand, was a stunning beauty, a consummate grifter and hustler in the truest sense.

If there were angles to exploit, Carrie would find one, and she made a fair living doing it. She'd been busted several times for theft, drug possession, and other crimes, but had always managed to stay out of jail. Just flashing those big blue cow eyes and blubbering how she was so sorry and would never do anything like that again. Judges and juries had always bought it hook, line, and sinker every time.

Carrie approached Michelle and chatted her up. "Hey, nice day. How're you doing? Mind if I sit with you? I'm Carrie. Hi!"

"Sure, have a seat. I'm Michelle. My friends call me Shelly." Carrie slid in next to her like a cat wanting Michelle's undivided attention; "So what are you doing, just hanging out here all by yourself?"

"Just trying to clear my head and deal with some shit," Michelle told her, murmured wistfully.

Like the seasoned hustler she was, Carrie picked up on it right away; "Well, maybe I can help. I'm a pretty good listener."

"I've got some medical things going on, that's all. Do you know where I can get some Oxy? My VA Doctors won't give me anymore. I'm really hurting!"

"You know what, Shelly, I might. I have a few sources. Are you sure that's all you're looking for, Shelly?" Carrie looked directly at her and gave Michelle a sly little smile and a suggestive grin.

"Well, I guess we'll just see what develops!" Michelle looked back at her with the same knowing little grin. Michelle's unexpected acquaintance with Carrie perked her right up, even a little excited about the possibilities. Carry asked her, "You live around here, Shelly?"

"Yeah, I have a place in Old Town."

"Wow! Very cool! You must be rich!"

"Naw, my grandparents left me the house. I just moved into it a little while ago after I retired from the Navy. I'm probably not going to stay there, though. Too expensive to keep it. My disability and pension payments won't even cover the taxes!"

"Oh-wow, what's the disability for?"

"I was a Navy Pilot and got hurt in a bad helicopter crash. It screwed me up a lot, so they retired me for medical reasons."

"I'll bet you were super good at it!"

"Meh, I guess so. My friend Ali was the best, though. We went down together in that crash." She whimpered a little, chin quivering, trying to control her feelings.

"Aw, I'm sorry if I upset you. Are you okay, Shelly?"

"Yeah, it just hurts a little. I don't know about her anymore. She doesn't live here. She's back in her hometown in Texas now."

Carrie cooed sweetly in her ear, "Shelly, can we go to your place and get better acquainted?" Michelle cocked her head around in amazement, did a double take, and realized what Carrie was suggesting. That little turn of events perked Michelle up at the thought of a new companion, especially one so much younger and so pretty.

"Sure, why not? I guess so," she replied with a saucy little inflection and a sly grin.

Had she been well and more aware, Michelle would have recognized Carrie's hustle immediately. In her current state of mind, this seductive exchange got to her and seemed like an absolute godsend. She was hurting, craving some companionship and attention from someone, anyone. They left the waterfront together and headed for Michelle's place.

"What about that, Oxy—can you get it for me, Carrie?"

"I think so, but wait until later. You're upset now. Let me see if I can make you feel better first!"

Michelle was hooked now on a real rollercoaster ride of desire, excited by Carrie's sweet, suggestive manner. It was just a game to Carrie, a scam she ran so well.

Esplanade on the Embarcadero San Diego, Ca

They both had a great time together for a few weeks, doing what new lovers do. Carrie had gotten her the Oxy she wanted, but at much higher prices than she'd paid for it. Carrie just skimmed off the top for herself. Michelle never questioned her, though; she craved Carrie's company and the dope.

Even though she was using way too much, Michelle was feeling better and enjoying life a little with Carrie until her sources ran dry; "Carrie, I need more Oxy. I've run out again."

"Wow, Shelly, I don't know if I can get anything else right now. My source told me last time I got you some that he's run out and can't supply any more for a while."

Michelle shrieked! "C'mon, Carrie, I need it! Please see if you can find some more somewhere else!"

She howled in a near-hysterical fit at the thought of not having any more OxyContin.

"Okay, okay! I'll go out later, make a few contacts, and see what I can find." Carrie was getting really tired of Michelle's constant whining for more pills. She knew Michelle was using way too much, but she figured, "Man, this needy bitch is getting under my skin. What the fuck? It's none of my business if she wants to kill herself with this junk."

She enjoyed living in Michelle's house and mooching off of her, though; "Beats the shit out of sleeping outside!" she told herself. Carrie always seemed to have a way to keep herself up. Even though she didn't have a regular place of her own, she had many male and female 'friends' who helped her out with that when she needed it. They took care of her, and she took care of them—more or less like she was doing with Michelle now.

Charlie Hansen

Sometime later, one of Michelle's friends, Charlie Hansen, short for Charlene from Michelle's VA support group, got seriously worried about her. Nobody had seen Michelle for a while. Charlie felt she needed to find out what was happening with her and if she was okay. She talked with Carlos earlier that day.

"We're all really worried about her, Carlos. Did you have any luck connecting with her?"

"Yes, she wasn't at home when I went to see her, so I went to where you guys said she enjoyed hanging out. I found her sitting at the esplanade. She was kind of surly and distant."

"It seemed like she was a little strung out, too. She didn't want to talk at all about anything. She was agitated about something and withdrawn. I tried but had no luck, so I just left a business card with her and told her how to contact me."

Charlie was disappointed and said, "Damn, I better check up on her. She hasn't come around in weeks. I hate to see her drift away like this."

After talking to Carlos, Charlie headed directly to Michelle's house to see if she was okay. Michelle didn't answer when Charlie knocked on the door and repeatedly rang the bell. Charlie feared the worst. She feared Michelle might be sick or if something was wrong or "worse."

She took a deep breath and slowly inhaled, as if to buck up and prepare for something terrible. She rang the doorbell again and waited a bit. No answer, so she rang again and rapped hard on the door, calling out, "Michelle, Michelle—It's Charlie; answer the door!"

Still nothing. She went to the back door and rapped hard on it several times. Still no answer. She checked the door, noticing it was unlocked. "Okay, that isn't good," she thought to herself. Charlie opened it, walked in, looked around, and thought, "Seems pretty normal, except it hasn't been cleaned for quite a while. What's that awful smell!"

Empty liquor bottles, beer cans, food wrappers, and trash lying around. Cigarette butts and marijuana roaches filled the ashtray on the coffee table. Michelle's cold, lifeless body was laid out on the couch. In stunned silence, Charlie stared for a moment at her lying there; "Oh my god, Michelle—No!" Panic-stricken, Charlie called 911 for help.

Paramedics and the San Diego police showed up quickly. Michelle was pronounced dead at the scene. The body was taken away to the county morgue. Charlie was dumbfounded, sick to her stomach, and grief-stricken over what she'd found and for Michelle.

Carrie was heading back to Michelle's place long about then, noticing the commotion. All the police and ambulance units were around in front of it. She knew right away what had gone down. "Oh shit, what happened? I'd better disappear!" That was it for Carrie and Michelle's brief affair. The "Oxy" she'd gotten for Michelle had been laced with fentanyl and had killed her.

Angela Brooks

A few days after the crash and the fight in Syria…

Angela knew her dad was off on another assignment with his Seal team guys somewhere in the Middle East. She didn't know precisely where and hadn't spoken with him for several weeks. He wasn't regularly in touch with anyone when he was gone like this.

Angie lived in his house in San Diego while attending college. He'd kept it mainly for her. He wasn't there much, anyway. His folks had left it to him when they passed. Lucky for him, that was after the divorce, or he likely would have lost it in the settlement with his ex-wife.

The house was quiet, as it usually was that day. Angie wasn't the type to party or have people over. It was a calm refuge. A place for her to hold on to family ties that were broken now by her parents splitting up—a quiet sentinel in the San Diego suburbs.

Now a college junior, she had settled in for the summer, comforted by its normalcy. She'd gotten used to how the house seemed to hold its breath whenever her dad was gone. It had been a few months now since his latest deployment. His absence always weighed heavily on her; always fearful the worst would happen.

Angie had been jogging in Balboa Park that morning, earbuds in, podcast streaming. She returned home around eleven, cheeks flushed with exertion from her run, hair pulled back into a messy ponytail. She pushed open the door, mind drifting with school out for the summer.

As she walked into the kitchen, she noticed an unusual car pull up in the driveway—a black sedan, slick and serious-looking. It was not the sort of vehicle usually seen in that neighborhood.

Two uniformed Navy officers got out, approached the house, and rang the front doorbell. A chill went through her. A shivering fear rose as she answered the door. The two officers stood there rigidly. Their expressions were solemn, their demeanor conveying a tense formality that made Angie catch her breath. She swallowed hard, her fingers gripping the doorknob.

Nervously, she asked: "May I help you?"

"Miss Angela Brooks?" the male officer asked, his voice gentle but clipped.

"Yes, that's me. What's wrong?" Angie replied, her voice trembling.

"May we come in?"

"Yes, what's going on, please?" fearing what was coming next.

"Would you mind if we all sat down and chatted?" the officer said, motioning towards the couch.

"Okay, but please, just tell me what's happened!" Her mind was racing now. The second officer, a woman with sharp features and a sympathetic gaze, sat beside her. Her presence felt like a shroud to Angie, wrapping the place in a heavy silence.

She noticed how the sunlight slanted through the window, casting long shadows across the floor, and how the wall clock's ticking grew louder with every passing second.

The man cleared his throat, his eyes never leaving hers. "Ahem, Ms. Banks, I'm Commander Adams, and this is Lieutenant-Commander Collins. We're from the Casualty Assistance Office at the Coronado Amphibious Base."

A lump formed in her throat, her heart pounding so loudly she thought it might drown out what she was sure was coming next. She nodded, though she could barely manage it.

"It's with the deepest regret that we have to inform you," the Commander continued, "that your father, Master Chief Banks, was killed in action in Syria. He was engaged in combat operations with a group of ISIS fighters, and—the words just blurred together as her vision tunneled.

The room seemed to spin, and the air thickened around her. She felt as though she were sinking; the floor falling away beneath her.

The Commander's voice was distant, muffled by the roaring silence in her ears. "No… no…" Angie's voice was a ragged little whimper, as if speaking louder might shatter the fragile reality she was trying to cling to. "Are… are you sure?"

Collins reached out, her hand resting gently on Angie's arm. "We understand this is incredibly difficult for you, Angela; that's your name, right? Angela?"

"Yes, yes, that's my name," Angie told her, whimpering, trying to keep control.

"We're here to support you and provide any help you need."

Angela's gaze drifted to the framed photographs on the mantle—a smiling father, a twinkle in his eye, the strength in his stance. He had always been her rock, her unwavering anchor, even during her parent's divorce. To hear that he was gone felt like the ground beneath her feet had fallen away.

Angie's breath came in shallow gasps; tears blurred her vision. The officers continued to speak. "Ms. Banks, your father, was a selfless hero. He helped save his men and two female Navy pilots who were flying a helicopter that was shot down by ISIS terrorists."

Collins tried to steady Angie, offering her condolences and other details about his death. "Ms. Banks, we're here to help. If you need anything or help we can offer, just ask. Is there anyone you'd like to contact? Do you have anyone who can stay with you now?"

Angie could hardly process Collin's words. She struggled to breathe. The air in the room seemed thin and oppressive, the silence overwhelming. She clutched her chest as if to hold her heart in place somehow. Collins put her arm around Angela and did her best to comfort her. "My mother and my brother; I need to contact them now, please!"

"Do you have them in your phone contacts?" Collins asked.

"Yes, please call my mother now!" Angie did her best to stay in control, but she was losing it now.

Collins tapped the contact under 'Mom'; it rang, and Collins handed it to her. Angie spoke in a painful, whimpering voice to her mother, trying to stay composed.

"Mama, can you please come by Daddy's house right away? A couple of officers from the base on Coronado are here. Daddy is gone! Okay, okay, hurry!" Angie ended the call by saying, "She'll be here in a while. She lives in San Clemente."

"No worries, Ms. Banks. We'll stay with you until she arrives."

A couple of hours later, Sharon, her mother, arrived. The two of them sat together while the Casualty Assistance Officers briefed Sharon. Sharon gave them their leave as they finished. "Thank you for your consideration and for staying with my daughter Angela until I arrived. Please leave us now."

"If you need anything, let us know, and we'll help. We'll also let you know when there are more details. Please accept our condolences for your family's loss."

Angie felt disoriented, numb, and raw all at once. A piercing emotional pain. They exchanged a final, compassionate look, and Collins gave her a soft nod before they walked out, leaving Angie with her mom and the weight of her grief.

Sharon wasn't so upset about it. She had left Beau a couple of years before amidst a bitter separation. The two had never come to terms with each other, but she tried hard to support her daughter now.

Beau's house, once Angie's sanctuary, now felt cold and empty. Angie sank to the floor, Sharon beside her, trying to comfort her daughter. Echoes of the two officers' words haunted the surrounding space.

Reaching for her dad's photograph, she held it close as the reality of his loss sank in. It was like an endless void stretching out in front of her. In that sunlit room, now surrounded by the shadows of happier times, Angie Banks grappled with the profound emptiness left by the loss of Master Chief Beauregard Banks, her dad, the man who had been her hero, her everything.

Chapter 12: Althea's Lament

Ali gets the bad news

2018...

Two years had gone by since the events surrounding the crash. Ali had recovered physically but still had a few problems. She'd learned to cope with it as well as anyone could. She still had trouble with her back and problems walking now and then, especially when it was cold weather. But she'd mostly gotten back to normal.

Ali hadn't heard from Michelle for a while. Today, her phone rang. It was Sam Nash. "Yo, this is Ali," she answered.

"Ali—It's Sam."

"Oh hey, Sam! How're you doing? Been a while, man! What's up?"

"Ali, I got a call from Scott Deering today, your old CO. He wants to talk to you about something. He said it was important but didn't have your contact info. He told me he'd remembered we were friends and asked if I had your number. I told him I'd check with you first before I gave it to him and see if it's all right with you if I give it to him."

"Sure, give him my number. Did he say what he wanted to talk to me about? "

"No, just that he needed to contact you."

"Okay, well, just give it to him. Did you get his contact number?"

"Yeah, I'll text it to you."

"Okay, thanks; I'll call him back myself. How did he know to contact you, anyway?"

"I'm not sure; I suppose he got it from Colonel Bradford. They're pretty tight buds."

They talked for a while longer. When they'd finished their conversation, she called Deering right away.

Deering answered her call after several rings: "Commander Deering."

"Commander Deering, it's Ali Cabrillo. I got a call from Sam Nash. He told me you wanted to talk to me about something?"

"Yes, hello, Althea; good to hear your voice again after so long. Are you busy?"

"Nice hearing from you, sir. I'll make time for you; what's up?"

"I need to see you. It's important."

"How would we do that? I'm in Texas. Where the heck are you?"

"I'm flying through San Antonio on my way to San Diego. I knew you lived nearby in San Marcos, so I thought I'd try to get in touch and see you."

"Well, sure, we can meet, but why? What's so important that you'd go so far out of your way like this?"

"I'll tell you when I see you. Can we meet somewhere for a coffee or something? It's important."

"There's this coffee shop nearby my place. I go there a lot. I can meet you there. How does one PM tomorrow afternoon work for you?"

"Sure, see you then."

"Okay, I'll text you the address."

Deering didn't want to tell Ali about Michelle over the phone. He knew it would be traumatic news and would upset her a lot, especially after all she'd been dealing with, trying to recover from surgeries, physical therapy, and her other problems.

They met at the coffee shop, where Ali had met Jean and Ranger. Many local vets hung out there for extended chat time and coffee. Ali greeted him with, "Hey, Commander, how have you been?"

"Oh, I'm okay, but just call me Scott; how are you getting along? I see you have a new companion."

"Yes, I do. Scott, meet Ranger; he's also a veteran with a combat disability too!"

"I see that. Nice to meet you, Ranger!" How are you two getting along?"

"We're still getting acquainted. He's great, though; Ranger is a real old soldier and a loyal friend. He's very protective of me, too. I like that. Ranger is my therapy dog. He's good for me. We take care of each other."

They sat at a table near the coffee shop's front window. Ranger by her side, as he always was.

"So why did you want to see me, Scott?"

He was uncomfortable now, considering the news he was bringing her. "I have some terrible news, Althea. I didn't want to tell you over the phone. I'm sorry to be the one to tell you this, Ali." She was getting nervous now, wondering what was coming next.

"Your friend Michelle from the squadron."

"Yes, what's wrong? What happened?" she asked nervously.

"Ali, she's gone."

Ali was stunned for a moment. "What, what happened?" stammering, trying to control herself, not being too successful at it.

"The police in San Diego aren't sure if she committed suicide or just overdosed. In either case, it was fentanyl poisoning that took her. As far as anyone knows, she was not doing very well."

"How, how do you even know about it?"

"They found a note in her place when the police came looking for her to do a welfare check. She'd been using a lot of street dope heavily. Apparently, her doctors cut off her pain meds. She asked that you be notified if anything happened to her."

Ali did her best to maintain her composure, but the breakdown came. Scott saw it coming and grabbed her arm. "Come on, let's get out of here." "Okay, please get me out of here, please—okay?"

"Sure, come on. Come on, Ranger, let's take care of your mom."

There was a park near the coffee shop where they'd met. Deering helped her to a bench, where they sat together for a while. Ali lost all composure, and the tears flowed. Scott tried to console her, but he had more to tell.

"Ali, she wanted you to attend the funeral. Michelle didn't have any close family. According to her note, you were her closest friend. She asked for you to be there for her."

"Oh god, how can I even think about that right now?" She thought she would cry, but the tears wouldn't come. The emotion faded. Then the realization crept in; Michelle was gone. At that point, she fell apart.

"I know; I'm so sorry, Ali, just let it out, just let it out."

Scott sat there with her and Ranger, trying to hold her up as best he could as she poured it out. The time they had worked together, the fun times, the moments of terror during the crash, and now this! "Why did she do it? What happened? Why, why, why?" she moaned.

Scott told her what he knew about it: "During their investigation, the San Diego PD found the drugs she'd been taking. The police report said it was faked OxyContin, laced with Fentanyl. The best they could figure out was Michelle got them from a street dealer, overdosed, and died from the junk.

The report also said it wasn't clear whether she intentionally meant to do it, but the result was the same. The police found what seemed like a suicide note. But they weren't sure if she'd intentionally overdosed herself with it or if it had just been accidental. Whatever it was, she died alone in her home in San Diego."

Deering added, "Ali, the funeral will be at the Fort Rosecrans national cemetery in a week. Many of your mutual friends from the squadron will be there for the both of you! Michelle had the forethought to set it up and made all the arrangements herself."

Ali got hold of herself a little and said, "Okay, Scott, I'll be there. I'll make the flight arrangements today."

"Bring Ranger, too. Everyone will be pleased to see you and him. Again, I'm so sorry to be the one to tell you this. I guess the only consolation is that she isn't suffering any longer. I'm going to have to go now, Ali. Are you going to be okay? Can you get home alright?"

"Yeah, I can handle it, I guess. See you there next week."

"Okay, then. I'll text you all the details as soon as I get everything. Take care of yourself, Ali, and that old soldier Ranger! See you there."

"Okay, Scott, thanks for telling me in person."

That was it; Scott Deering, her old squadron commander, left for San Diego. Ali sat there with Ranger for a while, gathering enough inner strength to get up and leave. She didn't hurry, though. It was a beautiful afternoon. The sunshine sent shafts of light through the oak trees while a soft, warm breeze made their leaves dance. It made her feel a bit more grounded as she struggled with the blow of Michelle being gone."

Michelle's Funeral

Fort Rosecrans National Cemetery, overlooking NAS North Island.

Fort Rosecrans National Cemetery, above Point Loma—Michelle's graveside service; a beautiful place overlooking Point Loma and the broad Pacific coast of California. Across the channel lay North Island Naval Air Station to the south. To the west, a wide panorama of the deep blue Pacific. It was far too beautiful a day for such a sad affair as a funeral. But then, most days in San Diego are beautiful.

Ali and the others were decked out in their best Dress Blues. Even though she'd been separated and resigned, she wanted to show it well, not for herself but for Michelle and the others attending. There was a pretty good turnout in true Navy fashion for a fallen shipmate.

Michelle's Navy buddies, Commander Deering, Sam Nash, and others she had served with, as well as a few of the aircraft maintenance and admin people from the cruise and Ranger, the war dog, of course, were there. There were also a few civilians Ali didn't know but assumed had somehow known Michelle.

The service began, and honors were rendered… Those in uniform stood at attention on the roadside next to the hearse. The pallbearers, also in full-dress blues, unloaded and carried Michelle's flag-draped coffin slowly as the honor guard performed a 'present arms' salute. Everyone saluted her smartly as her casket passed, then 'order arms.'

All present in uniform stood at attention, saluting her as she passed. The casket stopped at the graveside; the pallbearers placed her gently over the grave.

The Chaplain from the air station chapel gave the eulogy to the somber sound of taps as the flag that draped her coffin was folded. Since Michelle had no immediate family, she requested it be presented to Ali.

Deering officiated. He presented Ali with the flag and said: "On behalf of the President of the United States, the United States Navy, and a grateful nation, please accept this flag as a symbol of our appreciation for your loved one's honorable and dedicated service." A bell tolled three times; the rifle team fired a three-volley 'gun salute.'

Honors were rendered: Aviators from the squadron lined up to each 'pin,' a set of wings on her coffin—a time-honored tradition. Ali composed herself enough to do this one last thing for Michelle.

She was the last to pin wings of gold on Michelle's coffin. After the casket was lowered into the grave, everyone dispersed. Ali was spent. There were no tears left to shed for Michelle. She'd just become numb. "Why did it all happen? What was it all for?" Still no answers.

For today, though, she just wanted to return to her hotel but felt like she had to spend time with her old squadron mates before she left and reminisce about their times together. "Maybe it would help a little," she thought. "Maybe that will cheer me up."

Charlene, Michelle's friend from her VA support group, had been there, unofficially escorting Ali and Ranger while she was there for her funeral. The two walked towards Charlie's car, headed for the meetup, when Ali noticed a young woman approaching.

Angie Brooks walked up to her and asked her, "Excuse me, but are you Lieutenant-Commander Cabrillo, the pilot who flew with my dad the day he was killed?" Ali was overtaken and nearly lost it. "Oh, my god! Are you Master Chief Brooks' daughter, Angie?" Angie's lip quivered as she replied, "Yes, he was my dad. Can you tell me what happened that day?"

It was a heartbreaking scene. "Oh, honey, I'm so sorry about this. Come with us to the club at North Island. We're going to have a get-together. I'll tell you what your dad did for all of us that day. Some of the other people there knew him better than I did. You can talk to all of them if it helps."

"Thank you, I'd like that."

The Wake

It was a good-sized crowd; most people there had not seen each other for some time. Ali made the rounds, chatting with as many as she could. There was Master Chief Banks, who headed up the maintenance crews, and a few of his guys.

"Master Chief, hey! Good to see you again!"

"Hello, Ms. Cabrillo; so nice to see you again, Ma'am."

"Oh, call me Ali. I'm a civilian now."

"Very well, then. Call me Walt. I was very sorry to hear about Lieutenant Robbins. Things weren't the same after your crash. We all missed you both."

"I've so missed you guys, Walt; you don't know how much! It's been tough since then." Ali didn't mince words; she never did.

Long about this time, Deering joined in the conversation.

"Okay, what did I miss? Anything important?"

Ali introduced him to Angie. "Commander, this is Angela Brooks, Master Chief Brooks' daughter."

"Nice to meet you, Angela. My condolences about your dad. I didn't know him well, but I've heard nothing but good things about him."

"Thank you, Commander. I appreciate it. I miss him very much."

The chit-chat lasted a few hours, and the beer and cocktails flowed freely. The crowd broke up and started heading out. When it was finally over, and most everyone had left. Charlie, Ali, and Angie took off.

Charlie drove to Ali's hotel, pulled up in front of it, and asked, "Are you going to be okay, Ali? Do you want me to stay with you for a while?" Angie also volunteered. "I could stay with you too, Ma'am."

"Sure, that would be nice, but just call me Ali. We still have to have that talk about your dad." Ali struggled to grin a little, but didn't feel it. Charlie shut the car off. The three of them and Ranger went off to her room together. After a while, Charlie took her leave.

"Well, Ali, I'm going to say goodbye then. I hope I've made this a little easier for you. Angela, would you like a ride back to your car?"

Ali smiled appreciatively and told her, "You did, Charlie. You've been very helpful. I'm very grateful to you. Thank you."

Angela spoke up, replying to Charlie's offer, "Thanks for the offer, Charlie. I'll get an Uber. I still want to spend a little while longer with Ali."

"Yeah, I still need to tell you what happened."

"Oh, please, Ali! I want to know what happened. Maybe I'll be able to understand why he's gone better! How did you come by this poor crippled dog? He seems very attached to you."

"I adopted him from a place that finds homes for retired service animals. He was wounded and lost his front paw and part of his left leg in Syria, too. He was up for adoption and needed someone who understood his experiences to take care of him.

It turned out we both needed each other. Let's talk about what happened with your dad now."

"What happened? No one's ever told us anything. All they ever told us was he was killed in action in Syria, fighting some ISIS guys. Later, we got his Purple Heart and its award certificate. Through the mail! I made a Shadow box with all his stuff, including the Purple Heart."

Ali had to think about framing the story to help this young woman understand and be proud of her dad.

"Well, Angela, your dad saved me, Michelle, and a few others that day. He was knocked out of my helicopter after the ISIS rockets hit us."

"Even though he fell to the ground, he survived that and fought hard against those ISIS fighters trying to kill all of us."

"It got very brutal and overwhelming. He and a few of the other guys tried hard to protect us while a couple of them worked to get Michelle and me out of the wreck of that helicopter! Angie, your dad was a genuine hero."

"I wouldn't be here now if not for what he and those other Seal team guys and Marines did."

"Thank you for telling me all that, Ali. Now I miss him even more. But what are you going to do now? You're not in the Navy anymore, are you?"

"That's true, I'm not. I plan to return to school and study Psychotherapy. A friend, who was also wounded like me, did that. She now works with other veterans who suffer with PTSD. I've realized they need help. I feel like I owe it to Michelle, your dad, and the others who didn't make it to get involved and help."

"Wow, that's so selfless of you, Ali. You're a great person to do that for them."

"It'll help me too, Angie. Besides, I still have to take care of my buddy Ranger. Don't I, boy!" She grinned and looked down at the dog.

Ranger gave her that look, and big tail wags, thumping it on the floor.

"Well, I think it's time to head out. Has anyone seen Sam Nash?"

Deering answered the question, "I don't think he made it. I haven't seen him since everyone left the funeral service, either."

"Okay. Thank you all for coming today. I hope to see all of you again under better circumstances. Goodbye, folks!"

Then she left. And that was it. Ali already missed all of her friends but felt she just needed to get out of there after such a depressing experience as Michelle's funeral had been.

Haskins, the Attorney

The following morning, Ali's phone rang. "Hello."

Haskins replied, "Hello, is this Althea Cabrillo?"

"Yes, this is She. Who's this, please?"

"My name is Albert Haskins; I'm the attorney handling Michelle Robbin's estate. Is this a good time for you to chat?"

Confused, Ali asked, "Why do you want to talk to me?"

"You and she were friends, weren't you?"

"Yes, I suppose you could say that. Like sisters, really; why?"

"Well, as you're probably aware, Michelle had no immediate family."

"Yes, I was aware of that. What does that have to do with me?"

"Well, she stipulated in her will that everything was to go to you. Her house here in San Diego, her remaining money, and all her personal items. Unfortunately, what money she had was used to pay off her debts and the funeral expenses."

"Why would she have done that? We were friends and served together, nothing more than that."

"Well, I can't say with any certainty why she set it up this way, but that's what she wanted. Can you come to my office so we can finalize the transfer of her property and her belongings? You can do whatever you want with it after that."

"All right, I'll call you back and arrange a meeting with you. It'll take some time, though." Shocked again at her friend's actions, she thinks, "What the hell! Why would she ever do that?"

"Great, I look forward to meeting with you." Ali lied about not being able to go right away. She could have, but she was seriously surprised. "Why would Michelle have done this?" Ali was still really hurt over the whole thing: Michelle's death, the funeral, her issues, and now this.

After meeting with Haskins, the lawyer, Ali made her way to Michelle's place to take possession of what she'd just been given. What was left of Michelle's existence? She sat there going through Michelle's belongings and trying to understand. "Why, what happened? Had she killed herself?" Something Ali couldn't bring herself to do. Or was it just a tragic mistake, an overdose of junk?

This whole thing had only made it harder for her to cope with the stress and depression since the crash and resigning. She'd spent the last year and a half recovering from so much: a broken back and losing her leg, learning to walk on the prosthetic leg. Had it not been for Ranger, she might have done herself in, too. Ranger sat there next to her, put his right paw on her foot, and whined a little as if to show his sympathy for her sadness.

She went through everything, deciding what to keep and get rid of, looking for reasons. Sifting through all the papers Michelle left for her with the lawyer, she discovered why Michelle left everything to her.

Reviewing all the papers and documents, she came upon a manilla letter-sized envelope with her name handwritten in calligraphy. She opened it and started reading. Now, she realized Michelle's reasons. It was all clear now.

The script was also beautifully hand-written calligraphy:

"Ali

If you're reading this letter, I'm gone, and my attorney has already given you everything I owned. I suppose you would like to know why I gave it to you. As you know, I don't have any other family I would ever leave it to. Most of them have had nothing to do with me, so why should I give them anything?

You were my best, most precious friend. You helped me through many things I struggled with. I know I was a feckless bitch at times. That was just my nature, I suppose. I felt like it was the least I could do to leave you everything, since I couldn't give you what I wanted you to have. My heart, my body, myself.

I understand you were not like that, so I respected it for both our sakes and didn't pursue it since the service was such assholes about that kind of thing. All good. I understood why. I loved you, sweetheart: yes, I knew you on the inside, even though you were so strong, driven, and so good at what you did on the outside.

Goodbye, Ali: I'll be waiting for you on the other side, beloved!

Michelle "

Michelle's letter tore Ali up inside. She was never bi or a lesbian; she knew Michelle was, and she'd kept it to herself mostly, never realizing how deep it went for Michelle. She didn't understand all Michelle did to keep it inside to protect them both. "Oh, my god" was her only thought reading Michelle's letter. She wanted to cry about it even more now, but she had no more tears.

Sitting there processing it all, Ali realized many things at this revelation. She recalled thinking back on that time and their many conversations and experiences during that deployment. She felt stupid for not recognizing it.

Ranger whined a sympathetic whine for her, watching her with those big German Shepherd eyes. He always made her smile and feel a little better. It even helped now.

A Final Goodbye

Memorial Day…

Ali and Ranger returned to the cemetery at Fort Rosecrans before heading home. It was Memorial Day, appropriate for a last visit with Michelle. Walking amidst the forest of gravestones, she made her way to Michelle's grave, Ranger by her side. Ali kneeled and spoke at the headstone as if Michelle could hear what she was saying, her voice wavering and pained.

"Shell, I don't know why you did what you did or if it was intentional or not, or why you cared about me so much. I feel so guilty for not understanding the pain this must have been for you. I'm so sorry, sweetheart. I never realized; I never thought of you in that way."

Ali, Ranger by her side, sat with Michelle for several hours, reminiscing and reflecting on their time together and friendship. Ali had a much stronger personality than Michelle ever had—probably too strong. That was part of it; even though they were close, Michelle was her subordinate; she had a role to play—the role her job called for.

Kneeling there, looking at the cold white marble of her headstone, she reflected on the Purple Heart again, contemplating what she thought it meant. After all, she still felt it meant a person who'd survived the burning pain of wounds and losing friends and comrades in arms—feelings that are more painful and longer-lasting than physical wounds. But in reality, the award meant even more.

She'd brought Michelle's Purple Heart with her that day—the one she found in Michelle's house on the fireplace mantle and left it there for her. As she placed it, she said,

"Shell, I thought you should have this with you so people will see it and maybe understand. It means you were a loyal and loving friend who gave up everything for everyone else and for me! I felt it, too, for what it's worth now! I loved you too, baby, and I miss you terribly."

Parting recognition by Ali for Michelle's loyalty and devotion to her.

Ali and Ranger say goodbye to Michelle at Fort Rosecrans Cemetery on Memorial Day.

Cast of Characters

USS Guantanamo Bay

- Lieutenant-Commander Althea 'Ali' Luisa Cabrillo
- Lieutenant Michelle Robbins
- Major Samuel Nash—Ali's Freind and Old Fling
- Captain Jason Manning—Ship's Captain
- Commander Scott Deering—Squadron Commander
- Master Chief Walter Banks—Squadron Maintenance Chief
- Lieutenant-Commander Micheal Fox
- Petty Officer Joe Flanigan—Cabrillo's AWO weapons operator

Seal Team

- Master Chief Beauregard "Bo" Brooks—Seal Team OIC
- Angela Brooks—Brooks' Daughter
- Petty Officer Gilbert Greene—Seal Team Operator
- Petty Officer Faustino Garcia—Seal Team Operator
- Petty Officer Max Muldoon—Seal Team Operator
- Petty Officer Gregory Mallory—Seal Team Operator
- Petty Officer Tony Calabrese—Seal Team Operator
- Petty Officer Shane Sandoval—Seal Team Operator

Medevac Medical crew

- Warrant Officer Bryan Hayes—Medevac Flight Surgeon 'PA'
- HM2 Sheila Sanchez—Medevac med-tech

Ship's Medical Staff

- Commander Eric Slatter—Ship's Medical Officer and Surgeon
- HM2 Howie Burdick—Hospital Corpsman & Medical Attendant
- HM2 Sheila Sanchez—Hospital Corpsman

Sigonella Naval Hospital

- Captain J.D. Conway—Chief Orthopedic Surgeon
- Dr. Claudio De Luca—Hospital's Resident Anesthesiologist
- Dr. Angelo Modena—Chief of physical therapy

Bethesda—Walter Reed National Military Hospital

- Portia Radcliff—Head of Physical Therapy

Balboa Naval Hospital

- Captain Albert Smith—Medical Review Board Ranking Officer
- Dr. Carver—Resident Psychologist
- Commander Todd—Medical Review Board staff officer
- Commander Stevens—AirPac Staff Officer

San Marcos, Texas

- Colonel Jean Schaefer, USAF retired—Althea's Friend
- Ranger—The Retired War Dog

San Diego, California

- Lieutenant Colonel Samuel Nash–Still Ali's Friend
- Gunny Susan Barrett U.S.M.C.—Ali's Trainer at MCAS Miramar
- Albert Haskins—Michelle's Estate Attorney

- Carrie Lambert—Grifter, Michelle's Special Friend
- Charlene 'Charlie' Hansen—Michelle's support group friend
- Carlos—Therapist from the VA

Addendum

For all my fellow veterans, I've included a list of typical PTSD symptoms, hoping they will seek help or their families or friends will recognize these symptoms and encourage them to do so.

Symptoms

Symptoms are likely to develop within three months after the traumatic event; they may last for weeks, months, and sometimes even years. The symptoms are classified into four categories:

Re-experiencing

After the trauma, victims or the survivors may involuntarily replay the incident vividly, or it may appear in their dreams.

Avoidance and numbing

Survivors or victims may avoid places, people, events, or activities connected to the incident and shy away from talking or thinking about it. They may also experience fear, horror, anger, or shame and blame themselves for the incident. They may stop doing routine activities and become uninterested in their favorite things. They might isolate themselves from others and have negative thoughts about people and situations.

Changes in mood and cognition

After a traumatic incident, you might experience emotions like guilt, fear, horror, anger, or shame and blame yourself for the incident.

Arousal

A person might become irritable, have angry outbursts, show reckless behavior, or harm others or themselves.

Other symptoms include nightmares, pain, sweating, nausea, shivers, panic, extreme alertness, difficulty concentrating on basic things, being easily startled, numbness, and inability to remember the trauma or thoughts, feelings, or emotions.

Specific symptoms for women

Symptoms can vary regardless of gender. Some symptoms may be more commonly experienced by women soldiers with PTSD. They can include flashbacks, nightmares, intrusive thoughts, and emotional distress. Women may also experience symptoms related to their reproductive health, such as changes in menstrual cycles or difficulties with pregnancy.

Each person's experience with PTSD is unique, and it's always best to consult with a healthcare professional for a comprehensive understanding of PTSD symptoms in women soldiers.

https://www.medicalnewstoday.com/articles/va-ptsd-symptoms

Thoughts for my brother and sister veterans… As of this writing, the VA estimates there are an average of 22 veteran suicides each day. This doesn't have to be! Please don't suffer; get help to deal with it!

https://taskandpurpose.com/military-life/truth-22-veteran-suicides-day/

Some parting words of encouragement…

"The proper function of man is to live, not just to exist. I shall not waste my days trying to prolong them. I shall use my time."

Jack London—1906

I would add: Don't waste your days' suffering; pick yourself up and do what Soldiers, Sailors, Marines, and Airmen have always done and still do best…

"Adapt, Improvise, Overcome."

Glossary of Military and Navy Terms

- AK's: AK-47 Russian Assault Rifle.
- BMG: 50 Caliber Browning Machine Gun.
- Bulkhead: Wall or partition on a ship.
- CO: Commanding Officer.
- Deck: Floor on board a vessel.
- Four Oh (4.0): A perfect score, outstanding.
- Gedunk: Candy or goody shop on Navy ships.
- Hatch: Watertight door.
- Head: Bathroom, toilet space.
- Humvee: 'High Mobility Multipurpose Wheeled Vehicle'
- Knee knockers: Reinforcement steel on a bulkhead opening.
- LSE: Landing Signal Enlisted—a Petty Officer who gives directions via hand gestures to pilots.
- Mast: Disciplinary review by a Captain.
- MRAP vehicle: Mine-Resistant Ambush Protected combat vehicle.
- M4: A 5.56 caliber military carbine rifle.
- M433 eggs launcher: Forty-millimeter grenades for the M433 launcher.
- Overhead: Ceiling inside a ship.
- Pad Eye: Point on a deck to secure aircraft and cargo.
- POG: Acronym; 'Personnel Other than Government.'
- Port: Left side of a ship.
- Put in Hack: Restricted to one's quarters.
- Rack: Navy and Marine term for a cot or bed.
- RAG outfit: Replacement Air Group, a training or reserve squadron.
- Scuttle: A small round hatch in a bigger hatch or door.
- Scuttlebutt: Rumors, gossip, B.S., and a drinking water fountain (that often tastes like jet fuel!)
- Shit Can: Trash receptacle.
- Shove off: Leave now, get going.

- Squared Away: Perfect, as something should be.
- Starboard: Right Side of a ship.
- Stow: Put away, lockdown, secure.
- Turn To: Go to work and quit wasting time.
- XO: Executive Officer, second in command.

Made in the USA
Coppell, TX
09 June 2025